Synthesis Lectures on Engineering, Science, and Technology

The focus of this series is general topics, and applications about, and for, engineers and scientists on a wide array of applications, methods and advances. Most titles cover subjects such as professional development, education, and study skills, as well as basic introductory undergraduate material and other topics appropriate for a broader and less technical audience.

Angelo Luongo · Daniele Zulli ·
Manuel Ferretti · Francesco D'Annibale

Perturbation Methods and Nonlinear Phenomena

Applications to Continuous Mechanical Systems

Springer

Angelo Luongo
DICEAA
University of L'Aquila
L'Aquila, Italy

Manuel Ferretti
DICEAA
University of L'Aquila
L'Aquila, Italy

Daniele Zulli
DICEAA
University of L'Aquila
L'Aquila, Italy

Francesco D'Annibale
DICEAA
University of L'Aquila
L'Aquila, Italy

ISSN 2690-0300 ISSN 2690-0327 (electronic)
Synthesis Lectures on Engineering, Science, and Technology
ISBN 978-3-031-49396-6 ISBN 978-3-031-49397-3 (eBook)
https://doi.org/10.1007/978-3-031-49397-3

This Springer imprint is published by the registered company Springer Nature Switzerland AG
The registered company address is: Gewerbestrasse 11, 6330 Cham, Switzerland

Paper in this product is recyclable.

To my wife Fiorella, for her patience in enduring my silences and her encouragement to complete the work. (Angelo Luongo)

To my partner in life and one true love. (Daniele Zulli)

To the woman who instilled in me the power of perseverance. To my mother, Emilia. (Manuel Ferretti)

To Valeria, Gregorio and Bernardo for their unconditional love and support. My safe haven, my strength. (Francesco D'Annibale)

Preface

Perturbation methods (also said asymptotic methods) supply approximate solutions to weakly nonlinear algebraic or differential equations, containing a small parameter, even artificially introduced. Although there exist a number of different methods, all of them are aimed at reducing a nonlinear equation to a set of linear equations governed by the same operator, in which the known terms of the kth set only contains quantities already determined by the preceding k–1 sets. Thus, the equations can be solved in sequence, supplying successive contributions to series expansions describing the solution. In statics, the procedure leads to express nonlinear equilibrium positions; in dynamics, to obtain bifurcation equations governing the motion which develops on a finite-dimensional *invariant manifold* of small dimension (typically 1 or 2).

Perturbation methods are widely used in *Nonlinear Dynamics* to tackle several problems: internal resonances, primary, sub- or super-harmonic resonances, parametric excitation in principal, secondary or combination resonances. Several books by A. Nayfeh (1933–2017) present a large gallery of applications. Their use, in contrast, is less known in dynamic bifurcation problems, where the *Center Manifold Method*, in conjunction with the Normal Form Theory, is, still nowadays, more popular among the researchers, although many papers showed, in the last decades, that the same results can be more simply achieved by perturbation methods.

On the other side, *static perturbation methods* have represented the systematic tool to investigate buckling and postbuckling problems. They allow constructing bifurcated paths branching from a known fundamental path, which, in turn, may be built-up by asymptotic expansions. Examples of this approach, following pioneering studies by W. T. Koiter (1914–1997), have been extensively worked out in the literature, mainly by the British school in the 80s.

Perturbation methods have been more largely applied to discrete, rather than, continuous systems. It is not rare, yet, to find papers in which, after the continuous model has been formulated, a Galerkin discretization is performed, followed by a perturbation analysis. Such an approach, however, suffers from having to choose in advance the pattern of the displacement field, very often taken as a combination of linear modes, thus renouncing to capture the modification of the shape with the amplitude of the response (as predicted by the Center Manifold philosophy), or to describe it by a large number of degrees of freedom. A *direct approach*, therefore, in which the partial differential equations of motion are directly attacked, thus removing the limitation of the Galerkin method, received in the last years some attention, after some papers by Nayfeh and co-workers, although it has not been fully adopted in the international community.

This book is a contribution to spread the knowledge on applications of perturbation methods to *continuous systems* exhibiting several static and dynamic phenomena: nonlinear elastostatics, elastic buckling and postbuckling, nonlinear external resonances, parametric excitation, dynamic and static bifurcations of nonconservative systems. Although the methods differ among them to account for the different nature of the problem (homogeneous or non-homogeneous, autonomous or non-autonomous, governed by a non-singular or singular operator, self-adjoint or non-self-adjoint), the common aspects of the algorithms are highlighted, in an effort to present them in a unitary manner.

The book is intended for students of Ph.D. schools in Mechanical and Civil engineering. Only a basic knowledge of Mathematical Analysis and Algebra is required, as well as of Structural Mechanics and Dynamics (at least linear). Throughout the text, emphasis is not confined to computations, but a large space is left to the illustration of fundamental concepts of bifurcation analysis and nonlinear dynamics, aimed to give physical insights into phenomena.

L'Aquila, Italy
January 2024

Angelo Luongo
Daniele Zulli
Manuel Ferretti
Francesco D'Annibale

Contents

Introduction

1

1.1 Nonlinear Phenomena

Slender mechanical systems, when experience large displacements in the elastic range, manifest a number of interesting phenomena, in static as well as in dynamics, which cannot be explained by merely linear models. In contrast, they call for formulating nonlinear governing equations, and using proper computational tools for finding solutions.

Static Bifurcations

As a first example, let us consider a tubular beam of thin annular cross-section in large uniform bending. By assuming that the Navier solution for stresses is still valid in the nonlinear field, and due to the fact that the normal stresses act tangentially to the curved beam axis, transverse components arise, which cause flattening of the tube. The cross-section, therefore ovalizes, reducing its original inertia moment with respect the bending axis. By increasing the bending couple at the ends, the curvature increases more than linearly, till a limit point is reached, at which the beam loses stability, via a mechanism said of *fold bifurcation*. This phenomenon is known, in technical environment, as *Brazier effect* [4]. A similar mechanism is exhibited by the three-hinges low arch, which, when compressed, reduces its sag, and possibly jumps from the upper to the lower natural configuration, again when a limit point is reached [10, 17]. This occurrence is named *snap-through,* and, although entailing dynamic effects, it can still be studied in the static regime, if only the equilibrium points, and not the transient motions, are of interest.

Another static nonlinear phenomenon, very recurrent in applications, concerns the appearances of *branching points* (of fork or transcritical type [10, 17]). A celebrated example is offered by the *Euler beam*, that, when compressed by a gravitational force beyond a threshold

A. Luongo et al., *Perturbation Methods and Nonlinear Phenomena*,
Synthesis Lectures on Engineering, Science, and Technology,
https://doi.org/10.1007/978-3-031-49397-3_1

value, buckles, abandoning its initial straight configuration (which has become unstable), to assume a new bent configuration, which turns out to be stable [10, 17].

External and Parametric Resonances

When the nonlinear structures are studied in the dynamic field, new phenomena manifest themselves. The large undamped free oscillations of elastic systems are no more isochronous, i.e., their natural frequencies are no more independent of the oscillation amplitude, as it happens in the linear field. In contrast, the frequencies increase (hardening effect) or decrease (softening effect), compared to those of the infinitesimal motions, with a law which is at least quadratic in the amplitude. Moreover, if a harmonic external excitation acts on the structure with a frequency close to a natural one, i.e., if the excitation is in *primary resonance* with a natural mode of the structure, then the response does not diverge to infinity, as predicted by the linear theory, but is limited by the nonlinearity. As a further peculiarity, the bending toward the right or the left side of the classical amplitude-frequency curve, entails *multivalued responses*, and possible jumps, which cannot be described by the linear theory [14, 16].

Harmonic excitations acting on elastic structures also trigger forms of resonance which does not exist in the linear theory. They are called *sub-harmonic* and *super-harmonic* resonances, which manifest themselves when the natural frequency is an integer sub-multiple or multiple, respectively, of the external frequency. The phenomenon is due to the combination of a natural and the external frequency that, because of nonlinearities, create a new frequency which is in 1:1 resonance with the natural one [14, 16].

Another classic example concerns the *Bolotin beam*, which is a variant of the Euler beam, in which, however, the axial force harmonically depends on time [3]. For suitable combinations of the amplitude and frequency of the pulsating force, the beam loses stability and the response diverges to infinity (in the linear approximation) or approaches a limit cycle (in the nonlinear field). Such a phenomenon is called of *parametric excitation*, since the external action does not appear as a know term in the equation of motion, but rather enters the coefficients in the left hand side, modifying the intrinsic characteristics of the system [16].

Dynamic Bifurcations

A special attention deserves the analysis of elastic systems subject to *nonconservative loads*, as for example the well-known *Beck beam*, a cantilever compressed by a *follower force* applied at the tip [2, 7]. For such a problem, when the follower force exceeds a critical value, the rectilinear configuration loses stability via a *Hopf* (or dynamic) *bifurcation*, which entails the appearance of stable periodic motions (*limit cycles*) which did not exist at subcritical values of the load [19]. These cycles attract the trajectories, so that the beam, after having exhausted a transient, experiences periodic motions, sustained by the energy put into the system by the nonconservative action, whose amplitude is an increasing function of the intensity of the force.

Most of these phenomena will be browsed in this book. Other special occurrences, as internal resonances [16] or multiple bifurcations [8, 9, 11, 12] will not be addressed here, where the aim is to supply a basic, and possible simple, overview of the nonlinear behavior of continuous structures.

1.2 Perturbation *Versus* Numerical Methods

The analysis of continuous nonlinear systems is generally carried out by numerical methods, which, after discretization in space and time, usually provide almost exact solutions. On the other hand, there exist analytical methods, although approximate, which, *directly acting on the continuous model*, are able to qualitatively describe the main aspects of the phenomena under study, and often provide reasonable quantitative approximations. The analytical methods, moreover, supply parametric solutions, which do not require restarting the calculations for any set of data. Most importantly, they put in light some aspects of the solution, which is often not clearly readable when the approach is purely numeric. As a matter of fact, and according to the Center Manifold Theory [5, 6, 19, 20], when the system is close to a static or dynamic bifurcation, or it is in resonance with an external or parametric excitation, it behaves as a few degrees of freedom (DOF) system which, in principle, would not require a discretization in hundreds or thousand DOF.

In this book, the attention is focused on *perturbation methods*, i.e., asymptotic methods which provide analytical solutions for weakly nonlinear systems, expressed as small perturbations of linear solutions. There exist several perturbations methods. A detailed treatment is given in [15], and examples relevant to nonlinear dynamics are presented in [16]. Applications to dynamic bifurcation problems can be found, e.g., in [8, 9, 11, 12]. In this book, only the *Static Perturbation Method*[1] and the *Multiple Scale Method* (MSM) are illustrated, to solve static and dynamic problems, respectively.

1.3 A Sample Model: The Linear Elastic Beam on Nonlinear Winkler Soil

It will be seen, that the perturbation methods, able to deal with nonlinear elastic systems, work indifferently, and in the same way, on any systems. In order, however, to make the discussion less abstract, and, at the same time, to avoid repetitions, it is found convenient

[1] This falls in the categories of the *Straightforward Expansion Method* [15] (as used in the Sects. 2.2, 6.2 and Appendix B.2) or of the *Strained Parameter Method* (as used in the Sects. 2.3, 6.3 and Appendix B.3).

to introduce a *sample model*, as simple as possible, able to describe general methods and phenomena. This model is identified in a *linear* Euler-Bernoulli beam resting on *nonlinear* Winkler soil. Since all its nonlinearities are of algebraic type, mathematics is rendered simpler. The model is described in this section, and systematically used in the Chaps. 2–4 and Appendices A, B.

Sample Model

Let us consider a straight Euler-Bernoulli beam, linearly elastic, resting on a nonlinear (bilateral) Winkler elastic soil (Fig. 1.1). At this stage, in order to make the approach the easiest possible, we will consider that all the geometric nonlinearities of the beam are small and negligible with respect to those of the soil, so that the elementary theory of beam holds.

Let us assume that the beam is clamped at the left end A and free at the right end B, subjected to time-varying, transverse distributed gravitational forces (or dead loads) per unit length $\alpha p(s,t)$, and to a transverse gravitational force $\alpha P_B(t)$ applied at B, where α is a load multiplier, s is a material abscissa and t the time (Fig. 1.1a). Moreover, two forces G (longitudinal) and F (follower), are applied at the free end of the beam. In the reference configuration, they are both aligned with the beam axis and produce uniform compression; in the current configuration, described by the transverse displacement field $u\,(s,t)$, G moves with its application point by keeping its original direction, while F moves rotating of an angle $\theta_B \simeq u'_B$, by remaining aligned with the tangent at B to the beam axis. Occasionally, the force G will be considered made of a constant quota G_s (dead, or gravitational load) and of a quota $G_d \cos(\Omega t)$ harmonically varying in time, so that $G = G_s + G_d \cos(\Omega t)$. The gravitational force G_s is conservative (i.e., it admits a potential); the follower force F is nonconservative (since it spends a non-zero work on a cycle [2, 10]); the pulsating force $G_d \cos(\Omega t)$ is also non-conservative (since it depends on time).

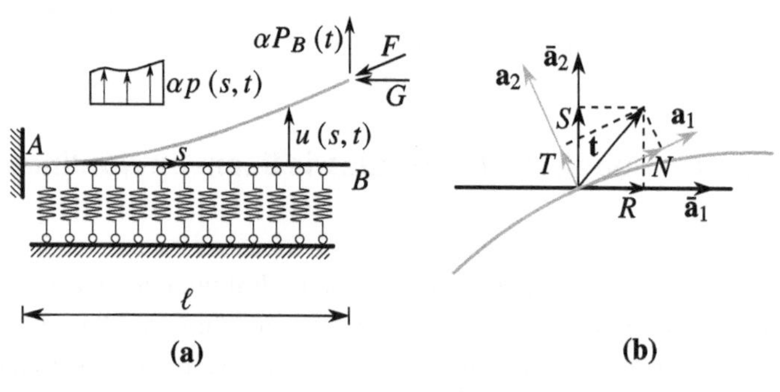

Fig. 1.1 Linear beam on nonlinear Winkler soil, compressed by a longitudinal force G and a tangential force F, acting simultaneously: **a** static scheme; **b** internal force **t** and its components (N, T) and (R, S) in the current and reference bases, respectively

According to the elementary theory of the inextensible, shear-undeformable and pre-stressed beam [13], the balance of the forces in the *adjacent configuration* (i.e., infinitely close to the reference configuration), requires that:

$$M'' - \left(N_0(s,t)u'\right)' = p_{ext}(s,t), \tag{1.1}$$

where $s \in [0, \ell]$ and $t \geq 0$. Moreover, $M(s) = EIu''$ is the bending moment, with EI the flexural stiffness of the beam and u'' the (linearized) curvature; $N_0(s,t)$ is the internal axial force acting in the reference configuration, constant in space and equal to $-(F+G)$; $p_{ext}(s,t)$ is the external force per unit length of beam, including inertia, viscous and elastic forces:

$$p_{ext} := \alpha p(s,t) - m\ddot{u} - c\dot{u} - r(u), \tag{1.2}$$

where m the mass per unit length, c is a damping coefficient, and $r(u)$ is the linear density of the elastic reaction exerted by the nonlinear Winkler soil; finally, a prime denotes space-differentiation and a dot time-differentiation.

Equilibrium at the free end requires that the internal force $\mathbf{t}_B$ and couple M_B at the boundary B equate the local incremental forces (i.e., the forces acting in the current state, subtracted the prestress forces). To write these conditions, we represent $\mathbf{t}$ both in the reference $(\bar{\mathbf{a}}_1, \bar{\mathbf{a}}_2)$ basis and in the current $(\mathbf{a}_1, \mathbf{a}_2)$ basis, by letting $\mathbf{t} = N\mathbf{a}_1 + T\mathbf{a}_2 = R\bar{\mathbf{a}}_1 + S\bar{\mathbf{a}}_2$ (Fig. 1.1b). Here N, T, as usual, denote the *normal* and *shear* forces (referred to the current basis), while R, S will be called the *longitudinal* and *transverse* internal forces (referred to the reference basis). Equilibrium at B then prescribes:

$$M_B = 0, \quad S_B = \alpha P_B(t) - Fu'_B, \tag{1.3}$$

in which $-Fu'_B$ is the transverse component of the follower load in the adjacent configuration. Note that no incremental forces act along $\bar{\mathbf{a}}_1$, since, in the adjacent configuration, $F\cos\theta_B \simeq F$; therefore, equilibrium of the prestressed state in the longitudinal direction is not disturbed. To express the transverse boundary condition in terms of displacements, it is worth noticing that $S_B = N_B \sin\theta_B + T_B \cos\theta_B$, or, by using linear kinematics, $S_B \simeq N_B u'_B + T_B$, with $N_B = -(G+F)$ and $T_B = -M'_B = -EIu'''_B$.

By collecting all the previous results, and appending geometric boundary conditions to be satisfied at the clamp A, the equations of motion finally read[2]:

[2] Note that the follower force does not enter into the boundary conditions. This circumstance is sometime explained by interpreting the last mechanical condition in Eq. 1.4 as a prescription to the shear force, namely $T_B = \alpha P_B + Gu'_B$, instead of considering the transverse force S_B, as we said. Although the two explanations appear to be equivalent (at first order, of course), our treatment is consistent with a variational formulation [13], in which the virtual work $S_B \delta u_B$ must vanish for any virtual displacement δu_B.

$$m\ddot{u} + c\dot{u} + EIu'''' + (F + G)\,u'' + r(u) = \alpha p(s,t), \tag{1.4a}$$

$$u\,(0,t) = 0, \quad u'\,(0,t) = 0, \tag{1.4b}$$

$$u''\,(\ell,t) = 0, \quad -EIu'''\,(\ell,t) - Gu'\,(\ell,t) = \alpha P_B(t), \tag{1.4c}$$

to be equipped with initial conditions $u(s,0) = u_0(s)$, $\dot{u}(s,0) = \dot{u}_0(s)$.

In the following, we will assume a polynomial series for the nonlinear reaction of the elastic soil, namely:

$$r(u) := \sum_{i=1}^{\infty} k_i u^i(s,t), \tag{1.5}$$

where k_i are known constants. If the soil behaves identically in compression ($u < 0$) and in traction ($u > 0$), i.e., if $r(-u) = -r(u)$, then only odd powers are contained in the series in Eq. 1.5. Since also the stresses of the beam possess this property, we will refer to this system as *symmetric*.[3] In the more general case, we will call it *non-symmetric system.*

Nondimensional Equations

By introducing the following nondimensional quantities:

$$\begin{aligned}
&\tilde{t} := \omega t, && \tilde{s} := \frac{s}{l}, && \tilde{u} := \frac{u}{l}, && \omega^2 := \frac{EI}{ml^4},\\
&\tilde{\Omega} := \frac{\Omega}{\omega}, && \xi := \frac{c\omega\ell^4}{EI}, && \tilde{k}_i := \frac{k_i\ell^{3+i}}{EI}, && \tilde{p} := \frac{p\ell^3}{EI},\\
&\tilde{P}_B := \frac{P_B\ell^2}{2EI}, && \mu := \frac{F\ell^2}{2EI}, && \nu_s := \frac{G_s\ell^2}{2EI}, && \nu_d := \frac{G_d\ell^2}{2EI},
\end{aligned} \tag{1.6}$$

where factors 2 have been introduced for later convenience, Eq. 1.4 are rendered nondimensional, as follows:

$$\ddot{u} + \xi\dot{u} + u'''' + 2(\mu + \nu_s + \nu_d\cos(\Omega t))\,u'' + \sum_{i=1}^{\infty} k_i u^i = \alpha p(s,t), \tag{1.7a}$$

$$u(0,t) = 0, \quad u'(0,t) = 0, \tag{1.7b}$$

$$u''(1,t) = 0, \quad -u'''(1,t) - 2\,(\nu_s + \nu_d\cos(\Omega t))\,u'(1,t) = \alpha P_B(t). \tag{1.7c}$$

Here tilde has been omitted and dots and primes now denote differentiation with respect to the nondimensional variables. These equations will be used in the first part of the book to illustrate perturbation methods and discuss mechanical phenomena.

[3] Note that an *antisymmetric* function describes a *symmetric* behaviour, according to the physical perception we have of the phenomenon.

1.4 Book Overview

The matter illustrated in this book is organized as follows. In the first part of the book (Chaps. 2–4) the simplest *sample system*, i.e., the *linear* beam resting on a *nonlinear* Winkler soil, introduced in Sect. 1.3, is used to illustrate *all* the phenomena to be discussed. In this model, the unique sources of nonlinearity are due to the soil, appearing in algebraic form, thus rendering the equations simple to be tackled. For the sample system, three classes of problems are studied:

- In the Chap. 2, the beam is analyzed in the static regime, to evaluate (i) the nonlinear elasto-static response and (ii) the buckling and postbuckling phenomenon induced by axial gravitational forces. Here, the static perturbation method is used, namely the *straigthforward expansion*, in the first problem, and the *strained parameter method* in the second problem. The methods furnish, in analytical form, the load-displacement curve emanating from the origin, as well the path branching from the bifurcation point.
- In the Chap. 3, the forced dynamics of the beam, excited by harmonic forces, is studied. First, transverse harmonic loads are considered, causing primary, sub-harmonic or super-harmonic resonances. Then, an axial pulsating gravitational force is considered, triggering parametric excitation. The *Multiple Scale perturbation Method* (MSM) is used to get the *bifurcation equation*, i.e., an ordinary differential equation which rules the slow dynamics of amplitude and phase of the mode involved in the resonance. The method supplies: (i) the frequency-response law for any specific external resonances, and (ii) the stability boundaries in the parameter plane, and the limit cycle from there emanating, in the parametric resonance case. Moreover, it allows detecting stability of the motions bifurcating from the trivial equilibrium.
- In the Chap. 4, the dynamic bifurcation suffered by the beam when loaded by a tangential (follower) force is studied, as well as the static bifurcation occurring when an additional gravitational force is applied. The MSM is used again to get bifurcation equations, from which the stability limits, the limit cycle in the dynamic case, and the branched path in the static cases, are determined. A stability analysis of the bifurcated motions is also carried out.

The second part of the book (Chaps. 5–7) is devoted to generalize concepts and methods to more complex structural elements, namely:

- In the Chap. 5 a *metamodel* and several specific models are introduced. The metamodel constitutes a paradigm for a large class of models to be studied; it is expressed in terms of abstract linear and nonlinear operators (algebraic, differential, integral). Several *minimal model* of nonlinear structures (taut string, suspended cable, beam with restrained or unrestrained longitudinal motions, taut membrane, plate) are derived. Here, only the main terms of the equations are retained, able to capture the mechanical behavior. More accurate models should be found in devoted books (e.g., beam and cables in [13], and plates in [1, 18]).

- In the Chap. 6 the perturbation methods, already introduced, are applied to the metamodel, for which different static and dynamic analyses are carried out. Here the attention is focused on the algorithm, and not on the model nor on the physics, so that the goal is to find the bifurcation equation.
- In the Chap. 7 a rich gallery of solved problems, relevant to different structural elements, is presented. The reader, after having gained experience in dealing with the sample system and the metamodel, is invited to solve problems related to various, and more realistic, structures, as described by the minimal models. Step-by-step procedures are illustrated, to help him/her to reach the goal, and the solution is provided.

The last Chap. 8 summarizes the main findings of the book, concerning models, algorithms and results. The Appendix A provides some mathematical tools (mainly the compatibility condition for non-homogeneous singular operators), needed to the perturbation analysis. The Appendix B re-illustrates the perturbation methods for discrete systems, to which the continuous model could be reduced via the Galerkin method.

Symbology and Conventions

The mathematical symbology is the standard one, according to which: the scalars are indicated with italic, Latin or Greek letters; matrices are indicated with uppercase bold; column matrices and vectors with lowercase bold. Rare exceptions are stressed in the text.

All equations are numbered, with numbering reinitialized in each chapter. For example, Eq. 1.10 is the Eq. 10 of Chap. 1. In the case of a reference to a package of equations: the reference to Eqs. 1.10 should be understood as the whole package; the reference to Eq. 1.10a or Eq. 1.10a, b indicates the first, or the first two equations of the package. The same convention applies to the figures, which are labeled in the format Fig. 1.10, and possibly divided into sub-figures Fig. 1.10a, b, ….

The bibliography refers almost exclusively to books, given the educational nature of the book. Very few articles are cited for their specificity. The bibliography is placed at the end of each chapter, listed in alphabetical order, and cited in the text at the proper place. It refers to the specific topics discussed in the chapter, and possibly repeated in other chapters.

References

1. Amabili, M.: Nonlinear vibrations and stability of shells and plates. Cambridge University Press, New York (2015)
2. Bolotin, V.V.: Nonconservative problems of the theory of elastic stability. Macmillan, London (1963)

3. Bolotin, V.V.: The dynamic stability of elastic systems. Holden Day, San Francisco (1964)
4. Brazier, L.G.: On the flexure of thin cylindrical shells and other "thin" sections. Proc. R. Soc. Lond. A **116**(773), 104–114 (1927)
5. Carr, J.: Applications of centre manifold theory. Springer, New York (1981)
6. Guckenheimer, J., Holmes, P.: Nonlinear oscillations, dynamical systems, and bifurcations of vector fields. Springer, New York (1983)
7. Leipholz, H.H.: Stability of Elastic Systems. Sijthoff & Noordhoff, Alphen aan den Rijn (1980)
8. Luongo, A., Di Egidio, A., Paolone, A.: Multiple time scale analysis for bifurcation from a multiple-zero eigenvalue. AIAA J. **41**(6), 1143–1150 (2003)
9. Luongo, A., D'Annibale, F.: Double zero bifurcation of non-linear viscoelastic beams under conservative and non-conservative loads. Int. J. Non-Linear Mech. **55**, 128–139 (2013)
10. Luongo, A., Ferretti, M., Di Nino, S.: Stability and Bifurcation of Structures: Statical and Dynamical Systems. Springer, Cham (2023)
11. Luongo, A., Paolone, A.: Perturbation methods for bifurcation analysis from multiple nonresonant complex eigenvalues. Nonlinear Dyn. **14**(3), 193–210 (1997)
12. Luongo, A., Paolone, A.: Multiple scale analysis for divergence-hopf bifurcation of imperfect symmetric systems. J. Sound Vib. **218**(3), 527–539 (1998)
13. Luongo, A., Zulli, D.: Mathematical Models of Beams and Cables. Wiley, New York (2013)
14. Lacarbonara, W.: Nonlinear Structural Mechanics: Theory, Dynamical Phenomena and Modeling. Springer Science & Business Media, New York (2013)
15. Nayfeh, A.H.: Perturbation Methods. Wiley, New York (1973)
16. Nayfeh, A.H., Mook, D.T.: Nonlinear Oscillations. Wiley, New York (1995)
17. Pignataro, M., Rizzi, N., Luongo, A.: Stability, Bifurcation and Postcritical Behaviour of Elastic Structures. Elsevier, Amsterdam (1990)
18. Timoshenko, S.P., Woinowsky-Krieger, S.: Theory of Plates and Shells. McGraw-Hill, New York (1959)
19. Troger, H., Steindl, A.: Nonlinear Stability and Bifurcation Theory: An Introduction for Engineers and Applied Scientists. Springer, Wien (1991)
20. Wiggins, S.: Introduction to Applied Nonlinear Dynamical Systems and Chaos. Springer, New York (1990)

2 Conservative Static Systems

2.1 Introduction

Static systems, when loaded by forces which increase with a load multiplier, exhibit a *nonlinear behavior*, i.e., suffer loss of proportionality between loads and displacements, which is peculiar of the linear theory. It is of interest to express analytically such a nonlinear response (at any material point) *vs* the load multiplier. The sample system introduced in the Chap. 1, namely the beam on the Winkler elastic soil, is used here to illustrate perturbation techniques able to describe the nonlinear behavior of structures falling in two classes of problems: (a) beam loaded by transverse forces (possibly exhibiting limit points), (b) beam axially loaded by a compression force (leading to *buckling*, in absence or in presence of small transverse loads acting as *imperfections*). Problems (a) are solved as an extrapolation from the origin, which calls for solving linear non-homogeneous equations governed by the same *non-singular* operator (the elastic stiffness). Problems (b) are solved as an extrapolation from an *unknown branching point*, which calls for solving: (i) a linear eigenvalue problem, and (ii) non-homogeneous equations governed by the same *singular* operator (the elastic plus geometric stiffness at the branching point). The former method is called the *straightforward expansion* (since it requires expanding only the displacements); the latter, the *strained parameter method* (since it requires expanding also the load).

A. Luongo et al., *Perturbation Methods and Nonlinear Phenomena*,
Synthesis Lectures on Engineering, Science, and Technology,
https://doi.org/10.1007/978-3-031-49397-3_2

2.2 Nonlinear Static Response

In this section, we tackle the nonlinear problem of elastic equilibrium.[1] We consider the sample model introduced in the Sect. 1.3, consisting of an unprestressed ($\mu = \nu_s = \nu_d = 0$) linear beam on nonlinear Winkler soil, carrying time-independent dead loads $\alpha p\,(s)\,,\ \alpha P_B$, in the domain and at the boundary, respectively. Since dependence on time disappears, i.e., $u = u\,(s)$, the nondimensional balance Eqs. 1.7 and boundary conditions reduce to:

$$u'''' + k_1 u + k_2 u^2 + k_3 u^3 + \cdots = \alpha p(s), \tag{2.1a}$$

$$u(0) = 0, \quad u'(0) = 0, \tag{2.1b}$$

$$u''(1) = 0, \quad -u'''(1) = \alpha P_B. \tag{2.1c}$$

We mainly refer to non-symmetric systems ($k_2 \neq 0$) in explaining the algorithmic aspects; then, we will discuss how to account for symmetry ($k_2 = 0$).

2.2.1 Non-symmetric Systems

The boundary value problem in Eqs. 2.1 is solved asymptotically, by applying the (static version of the) *straightforward expansion* method.

Rescaling

To apply the perturbation method, we first need to recognize the existence of a small, nondimensional, *perturbation parameter* $0 < \epsilon \ll 1$ in Eqs. 2.1, such that the equation can be solved in closed-form when ϵ tends to zero. Equations 2.1, indeed, do not contain such a parameter, and this is quite common in applications. However, we can overcome this lack by performing a suitable *rescaling* of the dependent variable. For example, if we rescale the displacement and the load multiplier as $u \to \epsilon \hat{u}$, $\alpha \to \epsilon \hat{\alpha}$, with $\hat{u} = \mathrm{O}(1), \hat{\alpha} = \mathrm{O}(1)$, and then we divide the equations by ϵ, we obtain (hat omitted):

$$u'''' + k_1 u + \epsilon k_2 u^2 + \epsilon^2 k_3 u^3 + \cdots = \alpha p(s), \tag{2.2a}$$

$$u(0) = 0, \quad u'(0) = 0, \tag{2.2b}$$

$$u''(1) = 0, \quad -u'''(1) = \alpha P_B, \tag{2.2c}$$

in which the perturbation parameter now appears. The rescaling expresses that we are interested in displacements and forces small of order ϵ, for which the quadratic terms are smaller than the linear ones, and the cubic terms are smaller than the quadratic ones, and so on. Of

[1] A discrete version of the problem is dealt with in the Sect. B.2 of the Appendix B.

course, since the perturbation parameter has been artificially introduced, it will be possible, at the end of the procedure, to reabsorb it, by performing the inverse rescaling.

Perturbation Equations

Referring to the rescaled field Eq. 2.2a, we observe that, when $\epsilon \to 0$, the equation becomes linear, and therefore easily solvable. Therefore we look at the solution of the (weakly) nonlinear problem as a (small) perturbation of that of its linear counterpart. To achieve this goal, according to the *straightforward expansion* method [5], we express the unknown terms as a series expansion of the perturbation parameter, namely:

$$u = u_0 + \epsilon u_1 + \epsilon^2 u_2 + \cdots , \tag{2.3}$$

in which $u_1 \equiv du/d\epsilon|_{\epsilon=0}$, $u_2 \equiv 1/2 d^2u/d\epsilon^2|_{\epsilon=0}$, ..., are unknowns to be determined. By substituting the series in the Eq. 2.2a, and collecting terms with same power of ϵ, it follows:

$$\begin{aligned} &\left(u_0'''' + k_1 u_0 - \alpha p(s)\right) + \epsilon \left(u_1'''' + k_1 u_1 + k_2 u_0^2\right) \\ &+\epsilon^2 \left(u_2'''' + k_1 u_2 + 2k_2 u_0 u_1 + k_3 u_0^3\right) + \cdots = 0. \end{aligned} \tag{2.4}$$

Now, we want this equation to be satisfied for *any* ϵ, and *not* for a specific value of it. Therefore, the coefficients of each power of ϵ in previous equation must vanish separately. Since the same must occur for the boundary conditions in Eqs. 2.2b, c, when the series expansion is substituted in them, the following independent problems are derived:

Order ϵ^0:

$$u_0'''' + k_1 u_0 = \alpha p(s), \tag{2.5a}$$

$$u_0(0) = 0, \quad u_0'(0) = 0, \tag{2.5b}$$

$$u_0''(1) = 0, \quad -u_0'''(1) = \alpha P_B. \tag{2.5c}$$

Order ϵ^1:

$$u_1'''' + k_1 u_1 = -k_2 u_0^2, \tag{2.6a}$$

$$u_1(0) = 0, \quad u_1'(0) = 0 \tag{2.6b}$$

$$u_1''(1) = 0, \quad -u_1'''(1) = 0. \tag{2.6c}$$

Order ϵ^2:

$$u_2'''' + k_1 u_2 = -2k_2 u_0 u_1 - k_3 u_0^3, \tag{2.7a}$$

$$u_2(0) = 0, \quad u_2'(0) = 0, \tag{2.7b}$$

$$u_2''(1) = 0, \quad -u_2'''(1) = 0. \tag{2.7c}$$

Equations 2.5–2.7 are called *perturbation equations*. They have important characteristics, that they share with all problems to be analyzed ahead, namely: (a) they are linear, governed by the same operator[2]; (b) the right-hand sides contain terms all evaluated at the lower orders. Therefore, the perturbation equations can be solved in chain.

Solution

From Eq. 2.5a, said *generating equation*, the *generating solution* u_0 is evaluated; when it is substituted in Eq. 2.6, u_1 is determined, and so on. Thus:

$$u_0 = \alpha\chi_0(s), \qquad u_1 = \alpha^2\chi_1(s), \qquad u_2 = \alpha^3\chi_2(s), \qquad \ldots, \tag{2.8}$$

where χ_m $(m = 0, 1, \ldots)$ satisfy the following problems:

$$\chi_m'''' + k_1\chi_m = \begin{cases} p(s), & \text{if } m = 0, \\ -k_2\chi_0^2(s), & \text{if } m = 1, \\ -2k_2\chi_0(s)\chi_1(s) - k_3\chi_0^3(s), & \text{if } m = 2, \end{cases} \tag{2.9a}$$

$$\chi_m(0) = 0, \quad \chi_m'(0) = 0, \tag{2.9b}$$

$$\chi_m''(1) = 0, \quad -\chi_m'''(1) = \begin{cases} P_B, & \text{if } m = 0, \\ 0, & \text{if } m = 1, 2. \end{cases} \tag{2.9c}$$

When the procedure is stopped, e.g., at ϵ^2order, the series in Eq. 2.3, with ϵ reabsorbed, furnishes the solution:

$$u = \alpha\chi_0(s) + \alpha^2\chi_1(s) + \alpha^3\chi_2(s) + \cdots. \tag{2.10}$$

Therefore, at a selected abscissa s^*, the displacement $u(s^*)$, the nondimensional bending moment $u''(s^*)$ and nondimensional shear force $-u'''(s^*)$, are algebraic curves $f_i(\alpha)$ in the multiplier α, which are tangent to the linear solution at the origin.

Remark 2.1 The differential operator in the left-hand side is *non-singular*, since the associated homogeneous equation, with homogeneous boundary conditions, admits only the trivial solution. Consequently, all the perturbation equations admit a unique solution. When this property holds, the procedure illustrated here works.

Remark 2.2 The perturbation equations are susceptible of the following mechanical interpretation. The load is first applied to the linear structure, and the linear response u_0 is evaluated. Then, the quadratic nonlinearities, evaluated at u_0, are applied *as an incremental load* to the same linear structure, and an incremental displacement u_1 is determined.

[2] It is called the *tangent stiffness* operator.

Hence, the cubic nonlinearities, evaluated at u_0, and the quadratic, updated by u_1, are in turn applied, as incremental loads, to the linear structure, and the relevant incremental response u_2 is computed. When the procedure is truncated, the overall response is the sum of the incremental responses.

2.2.2 Symmetric Systems

A special, but important, case occurs when the soil behaves symmetrically, i.e., when $k_2 = 0$. Since Eq. 2.6a becomes homogeneous, and boundary conditions are also homogeneous, then $u_1 \equiv 0$. It is easy to check that $u_i \equiv 0, i = 1, 3, \ldots$. Symmetric systems are very frequently encountered, so that it is useful to take advantage from this circumstance, by omitting all the odd terms in the series expansion, that therefore reads $u = u_0 + \epsilon^2 u_2 + \epsilon^4 u_4 + \cdots$. This suggests to rename the old ϵ^2-parameter as a new ϵ-parameter, by proceeding, more effectively, as follows: (a) first, perform the rescaling $u \to \epsilon^{1/2}\hat{u}$, $\alpha \to \epsilon^{1/2}\hat{\alpha}$, which transforms Eq. 2.1a (with $k_2 = k_4 = \cdots = 0$), after division by $\epsilon^{1/2}$, into:

$$u'''' + k_1 u + \epsilon k_3 u^3 + \cdots = \alpha p(s); \tag{2.11}$$

(b) then, use again the series in Eq. 2.3, and obtain the (now all nontrivial) perturbation equations:

Order ϵ^0:

$$u_0'''' + k_1 u_0 = \alpha p(s), \tag{2.12a}$$

$$u_0(0) = 0, \quad u_0'(0) = 0 \tag{2.12b}$$

$$u_0''(1) = 0, \quad -u_0'''(1) = \alpha P_B. \tag{2.12c}$$

Order ϵ^1:

$$u_1'''' + k_1 u_1 = -k_3 u_0^3, \tag{2.13a}$$

$$u_1(0) = 0, \quad u_1'(0) = 0, \tag{2.13b}$$

$$u_1''(1) = 0, \quad -u_1'''(1) = 0. \tag{2.13c}$$

The solution, therefore reads:

$$u = \alpha \chi_0(s) + \alpha^3 \chi_1(s) + \cdots, \tag{2.14}$$

with:

$$\chi_m'''' + k_1\chi_m = \begin{cases} p\,(s)\,, & \text{if } m = 0, \\ -k_3\chi_0^3(s), & \text{if } m = 1, \end{cases} \tag{2.15a}$$

$$\chi_m\,(0) = 0, \quad \chi_m'\,(0) = 0, \tag{2.15b}$$

$$\chi_m''\,(1) = 0, \quad -\chi_m'''\,(1) = \begin{cases} P_B, & \text{if } m = 0, \\ 0, & \text{if } m = 1. \end{cases} \tag{2.15c}$$

2.3 Buckling and Postbuckling

In this section, we carry out the buckling analysis of the linear beam on nonlinear elastic soil.[3] Let us consider a clamped-free beam, prestressed by a nondimensional dead load ν. The governing equations are derived from Eq. 1.7, by letting $\nu_s = \nu$, $\nu_d = \mu = \alpha = 0$, and accounting for independence of time, thus obtaining:

$$u'''' + 2\nu u'' + k_1 u + k_2 u^2 + k_3 u^3 + \cdots = 0, \tag{2.16a}$$

$$u(0) = 0, \quad u'(0) = 0, \tag{2.16b}$$

$$u''(1) = 0, \quad -u'''\,(1) - 2\nu u'\,(1) = 0. \tag{2.16c}$$

These equations, being homogeneous, admit the trivial solution $u \equiv 0$; we want to find if nontrivial solutions exist too, and to determine, if any, in which range of the load parameter ν they are defined. We will first consider the general non-symmetric case ($k_2 \neq 0$), then we will study the symmetric case ($k_2 = 0$). At the end of the section, we will shortly address the problem of the 'imperfections'.

2.3.1 Non-symmetric Systems

The boundary value problem in Eq. 2.16 is solved asymptotically, by applying the (static version of the) *strained parameter* method.

Rescaling

To study Eq. 2.16, we first rescale the variable as $u \to \epsilon\hat{u}$ and divide the equations by ϵ (we omit the hats), to obtain:

---

[3] A discrete version of the problem is dealt with in the Sect. B.3 of the Appendix B.

$$u'''' + 2\nu u'' + k_1 u + \epsilon k_2 u^2 + \epsilon^2 k_3 u^3 + \cdots = 0, \tag{2.17a}$$

$$u(0) = 0, \quad u'(0) = 0, \tag{2.17b}$$

$$u''(1) = 0, \quad -u'''(1) - 2\nu u'(1) = 0. \tag{2.17c}$$

Perturbation Equations

Then, we have to expand the variable. However, differently from statics, the load ν is here unknown, and therefore we also need to expand it around an unknown value ν_0. For this reason the perturbation method is called of the *strained parameter method* [5]. We will see that this task plays a fundamental role in pursuing our goal. According to this strategy we put:

$$u = u_0 + \epsilon u_1 + \epsilon^2 u_2 + \cdots, \tag{2.18a}$$

$$\nu = \nu_0 + \epsilon \nu_1 + \epsilon^2 \nu_2 + \cdots, \tag{2.18b}$$

in which all the coefficients in the two expressions are unknown. By substituting the series expansions of Eq. 2.18 in the rescaled equilibrium Eqs. 2.17, and equating separately to zero the terms with the same power of ϵ, we get the perturbation equations with the relevant boundary conditions:

Order ϵ^0 :

$$u_0'''' + 2\nu_0 u_0'' + k_1 u_0 = 0, \tag{2.19a}$$

$$u_0(0) = 0, \quad u_0'(0) = 0, \quad u_0''(1) = 0, \tag{2.19b}$$

$$-u_0'''(1) - 2\nu_0 u_0'(1) = 0. \tag{2.19c}$$

Order ϵ^1 :

$$u_1'''' + 2\nu_0 u_1'' + k_1 u_1 = -2\nu_1 u_0'' - k_2 u_0^2, \tag{2.20a}$$

$$u_1(0) = 0, \quad u_1'(0) = 0, \quad u_1''(1) = 0, \tag{2.20b}$$

$$-u_1'''(1) - 2\nu_0 u_1'(1) = 2\nu_1 u_0'(1). \tag{2.20c}$$

Order ϵ^2 :

$$u_2'''' + 2\nu_0 u_2'' + k_1 u_2 = -2\nu_2 u_0'' - 2\nu_1 u_1'' - 2k_2 u_0 u_1 - k_3 u_0^3, \tag{2.21a}$$

$$u_2(0) = 0, \quad u_2'(0) = 0, \quad u_2''(1) = 0, \tag{2.21b}$$

$$-u_2'''(1) - 2\nu_0 u_2'(1) = 2\nu_2 u_0'(1) + 2\nu_1 u_1'(1). \tag{2.21c}$$

It should be noticed that, differently from the static problem, the generating equation, and boundary conditions, are now homogeneous, while all the successive equations, and boundary conditions, are non-homogeneous. In order for the generating equation to admit a non-trivial solution, we have to render singular the differential operator by a suitable choice of the free parameter ν_0. However, this entails that even the higher order perturbation equations are singular, and therefore compatibility conditions have to be enforced at each steps (see the Appendix A). These conditions can actually be satisfied, since we have the free parameters $\nu_1,\ \nu_2, \ldots$ still available, whose role in the algorithm is, therefore, essential.

Normalization Condition

The appearance of arbitrary constants related to the singular problem, calls for introducing a *normalization condition*, e.g., $u(1) = a$, where a is an amplitude. From this condition, by using the series expansion Eq. 2.18a, several normalization conditions follow, namely:

$$u_0(1) = a, \quad u_1(1) = 0, \quad u_2(1) = 0, \quad \ldots, \tag{2.22}$$

each joint to the perturbation Eqs. 2.19–2.21. The meaning of a can easily been explained. If we think to it as a prefixed value, we are prescribing the displacement at $s = 1$, and we are asking to ourselves: (a) which is the shape $u(s)$ assumed by the beam when its tip is displaced by a?; (b) which is the prestress ν in equilibrium with this configuration? If, in contrast, we think a as a free control parameter allowed to vary in a certain range, then we are describing a load path, relating displacement and prestress.

Solution

Let us start by analyzing the generating problem in Eqs. 2.19. This is a *boundary eigenvalue problem* in the eigenvalue ν_0, which admits infinite roots (assumed distinct) $\nu_{c_1} < \nu_{c_2} < \cdots$, said *critical*, or *bifurcation values* (also referred to as *Eulerian* critical loads). Associated with these eigenvalues, there exist a set of infinite (right) eigenvectors, or (buckling) modes, $\phi_1(s), \phi_2(s), \ldots$. Since the problem is self-adjoint (being $\mu = 0$), the homogeneous adjoint problem admits solutions $\psi_k(s) \equiv \phi_k(s)$, said left eigenvectors (see the Appendix A).

One is usually interested just in the lowest eigenvalue $\nu_c := \nu_{c_1}$, said *the* critical load, and in the associated *critical mode* $\phi_c(s) := \phi_1(s)$; therefore, a non-trivial solution of Eq. 2.19a reads:

$$u_0 = a\phi_c(s), \qquad \nu_0 = \nu_c, \tag{2.23}$$

where $\phi_c(1) = 1$, by virtue of the first of Eq. 2.22. Once the previous solution is substituted in the ϵ order perturbation Eqs. 2.20, these latter become:

$$u_1'''' + 2\nu_c u_1'' + k_1 u_1 = -2a\nu_1\phi_c'' - a^2 k_2 \phi_c^2, \tag{2.24a}$$

$$u_1(0) = 0, \quad u_1'(0) = 0, \quad u_1''(1) = 0, \tag{2.24b}$$

$$-u_1'''(1) - 2\nu_c u_1'(1) = 2a\nu_1\phi_c'(1). \tag{2.24c}$$

These equations cannot be solved for arbitrary values of the 'known term', since the differential operator is singular at $\nu_0 = \nu_c$. To ensure solvability, we have to enforce the compatibility condition discussed in the Appendix A. By recalling Eq. A.23, and that $\psi_c \equiv \phi_c$, the condition reads:

$$-\int_0^1 \phi_c \left(2a\nu_1\phi_c'' + a^2 k_2 \phi_c^2\right) \mathrm{d}s + 2a\nu_1\phi_c(1)\phi_c'(1) = 0. \tag{2.25}$$

This is an equation for the ν_1 unknown, from which, after integration by parts and accounting for $\phi_c(0) = 0$, it follows:

$$\nu_1 = C_1 a, \tag{2.26}$$

where:

$$C_1 := \frac{1}{2} k_2 \frac{\int_0^1 \phi_c^3 \mathrm{d}s}{\int_0^1 \phi_c'^2 \mathrm{d}s}. \tag{2.27}$$

Equation 2.24 can now be solved, to give the solution:

$$u_1 = a^2 \chi_1(s), \tag{2.28}$$

where $\chi_1(s)$ solves the singular problem:

$$\chi_1'''' + 2\nu_c \chi_1'' + k_1 \chi_1 = -2C_1\phi_c'' - k_2\phi_c^2, \tag{2.29a}$$

$$\chi_1(0) = 0, \quad \chi_1'(0) = 0, \quad \chi_1''(1) = 0, \tag{2.29b}$$

$$-\chi_1'''(1) - 2\nu_c \chi_1'(1) = 2C_1\phi_c'(1). \tag{2.29c}$$

To render the solution *unique*, the normalization condition $\chi_1(1) = 0$ is used, which follows from the second of Eqs. 2.22.

By going one step further, the ϵ^2-perturbation Eqs. 2.21 must be solved. With the results so far achieved, they read:

$$u_2'''' + 2\nu_c u_2'' + k_1 u_2 = -2a\nu_2\phi_c'' - 2a^3 C_1 \chi_1'' - 2a^3 k_2 \phi_c \chi_1 - a^3 k_3 \phi_c^3, \tag{2.30a}$$

$$u_2(0) = 0, \quad u_2'(0) = 0, \quad u_2''(1) = 0, \tag{2.30b}$$

$$-u_2'''(1) - 2\nu_0 u_2'(1) = 2a\nu_2\phi_c'(1) + 2a^3 C_1 \chi_1'(1). \tag{2.30c}$$

Again, since these equations are singular, they call for a compatibility condition, which reads:

$$\begin{aligned}&-\int_0^1 \phi_c \left(2a\nu_2\phi_c'' + 2a^3 C_1\chi_1'' + 2a^3 k_2\phi_c\chi_1 + a^3 k_3\phi_c^3\right) \mathrm{d}s \\ &+2\phi_c(1)\left(a\nu_2\phi_c'(1) + a^3 C_1\chi_1'(1)\right) = 0,\end{aligned} \tag{2.31}$$

from which ν_2 is evaluated as:

$$\nu_2 = C_2 a^2, \tag{2.32}$$

where:

$$C_2 := \frac{1}{2}\frac{2k_2\int_0^1\phi_c^2\chi_1\mathrm{d}s + k_3\int_0^1\phi_c^4\mathrm{d}s - 2C_1\int_0^1\phi_c'\chi_1'\mathrm{d}s}{\int_0^1\phi_c'^2\mathrm{d}s}. \tag{2.33}$$

Solution to Eqs. 2.30 could also be found in the form $u_2 = a^3\chi_2(s)$, but we will not further pursue it, since, usually, a solution truncated at this order is sufficient, and the procedure should be now clear to the reader.

By summarizing the results, the series in Eqs. 2.18, together with Eqs. 2.23, 2.28, 2.26 and 2.32, give:

$$u = a\phi_c(s) + a^2\chi_1(s) + \cdots, \tag{2.34a}$$

$$\nu = \nu_c + C_1 a + C_2 a^2 + \cdots, \tag{2.34b}$$

in which the perturbation parameter has been reabsorbed. Equation 2.34 describe, in parametric form, the non-trivial load-displacement relation, with $a \equiv u(1)$ being the parameter.[4]

Bifurcation Diagram

Let us now describe the beam behavior on the (ν, a)-plane (see Fig. 2.1, where only the leading term in Eq. 2.34b is represented). The ν-axis represents the trivial solution, along which the beam remains straight. Equation 2.34b describes the non-trivial solution, along which the beam buckles in a pattern established by Eq. 2.34a. The trivial branch is said *fundamental path*, the non-trivial the *bifurcated path*. The point $C := (\nu = \nu_c,\ a = 0)$ at which they mutually intersect, is called a *branch*, or *bifurcation point*. Here a *bifurcation* is said to occur, since the dynamic (static) behavior of the system abruptly changes, when the parameter ν quasi-statically crosses it; therefore ν is said a *bifurcation parameter*. The whole plot is also said a *bifurcation diagram*. Stability analysis of the branches requires accounting for dynamic effects, that will be introduced later (see the Sect. 4.3).[5] Results are displayed

[4] In fact, it is customary in the buckling literature (see, e.g., [6]), to not introduce a, but using ϵ as a parameter. However, we prefer this approach, aimed to preserve similarities with the dynamic problem to be tackled ahead.

[5] Indeed, for conservative systems like this, dynamic effects can be ignored, and reference made to the Total Potential Energy criterion, as usually done in buckling theory (see, e.g., [4, 6]). In this book,

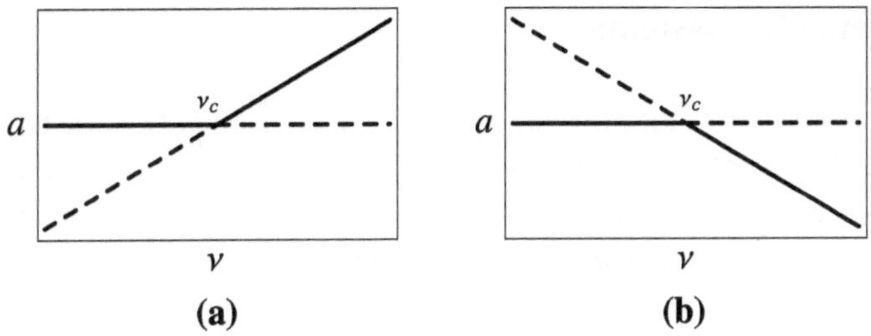

Fig. 2.1 Bifurcation diagram (transcritical) for the non-symmetric beam, axially prestressed by a dead load: **a** $C_1 > 0$, **b** $C_1 < 0$; first-order solution

in Fig. 2.1, where stable branches are continuous, and unstable are dashed. It is seen that an *exchange of stability* occurs at the bifurcation point, where the fundamental paths loses its own, irrespective of the sign of C_1. Such a bifurcation is known as *transcritical bifurcation*, since stable equilibria exist on both sides of C.

Remark 2.3 It should be noticed that the most important contribution to the description of the bifurcated path (Eq. 2.34b) is linear in the amplitude (associated with quadratic non-linearities, see Eq. 2.26). As a matter of fact, the term depending on the squared amplitude (produced both by quadratic and cubic nonlinearities, see Eqs. 2.32, 2.33), does not qualitatively affect the path, if the load is sufficiently close to the bifurcation value. Therefore, in dealing with non-symmetric systems, it is customary to truncate the analysis at the ϵ order, thus approximating the bifurcated path by its tangent at bifurcation. According to this approximation, $u = a\phi_c(s)$ and $\nu = \nu_c + C_1 a$ should be taken, from which the parameter a can be eliminated to obtain $u = \phi(s)\,(\nu - \nu_c)\,/C_1$. This is called a *first order perturbation solution* (as opposite to the second order solution of Eq. 2.34), to indicate that just the first nontrivial information has been taken into account in the algorithm.

Remark 2.4 It is worth noticing that quadratic nonlinearities produce a *change in the shape* of the buckling pattern, through the function $\chi_1(s)$. Therefore, Eq. 2.34 describe a deflection pattern which evolves (not only in magnitude) with the load intensity. The alteration $\chi_1(s)$ of the shape, however, is 'slave' of the 'master' buckling mode $\phi_c(s)$ (remember Eq. 2.29), so that the evolution problem remains one-dimensional, encompassed by the amplitude-load relationship. This is main idea of the Center Manifold Theorem [1, 3, 7, 8], which guides the bifurcation analysis.

---

however, in which we analyze both static and dynamic systems, it seems more elegant to use the same (and more general) tool for stability investigation.

2.3.2 Symmetric Systems

When $k_2 = k_4 = \cdots = 0$, the previous equations strongly simplify, since the compatibility condition Eq. 2.26 supplies $\nu_1 = 0$. Consequently, the ϵ order perturbation Eqs. 2.20 become homogeneous and, under the normalization condition $u_1(1) = 0$, furnish $u_1(s) \equiv 0$, i.e., they degenerate in a trivial step in the perturbation process. Similarly to what observed in nonlinear statics, it could be checked that all the odd terms of the series expansion of Eqs. 2.18a, b vanish. This property can be easily explained by the fact that, because of the symmetry, replacing ϵ by $-\epsilon$ in the series, the displacements $u = \epsilon\hat{u}$ must change sign, while the axial force ν must keep its sign. One can take advantage of this consideration, by renaming the old ϵ^2 as a new ϵ, by operating as follows. First, the rescaling $u \to \epsilon^{1/2}\hat{u}$ is used, which transforms Eqs. 2.16 into:

$$u'''' + 2\nu u'' + k_1 u + \epsilon k_3 u^3 + \cdots = 0, \tag{2.35a}$$

$$u(0) = 0, \quad u'(0) = 0, \quad u''(1) = 0, \tag{2.35b}$$

$$-u'''(1) - 2\nu u'(1) = 0. \tag{2.35c}$$

Then, the series expansions Eqs. 2.18 are substituted in it, leading to the following perturbation equations:

Order ϵ^0:

$$u_0'''' + 2\nu_0 u_0'' + k_1 u_0 = 0, \tag{2.36a}$$

$$u_0(0) = 0, \quad u_0'(0) = 0, \quad u_0''(1) = 0, \tag{2.36b}$$

$$-u_0'''(1) - 2\nu_0 u_0'(1) = 0. \tag{2.36c}$$

Order ϵ^1:

$$u_1'''' + 2\nu_0 u_1'' + k_1 u_1 = -2\nu_1 u_0'' - k_3 u_0^3, \tag{2.37a}$$

$$u_1(0) = 0, \quad u_1'(0) = 0, \quad u_1''(1) = 0, \tag{2.37b}$$

$$-u_1'''(1) - 2\nu_0 u_1'(1) = 2\nu_1 u_0'(1), \tag{2.37c}$$

where we confined ourselves to the first order.

The generating system, Eqs. 2.36, admits the solution $u_0 = a\phi_c$, $\nu_0 = \nu_c$; compatibility for Eq. 2.37 reads:

$$-\int_0^1 \phi_c \left(2a\nu_1\phi_c'' + a^3 k_3 \phi_c^3\right) \mathrm{d}s + 2\phi_c(1) a\nu_1 \phi_c'(1) = 0, \tag{2.38}$$

from which:

$$\nu_1 = C_2 a^2 \tag{2.39}$$

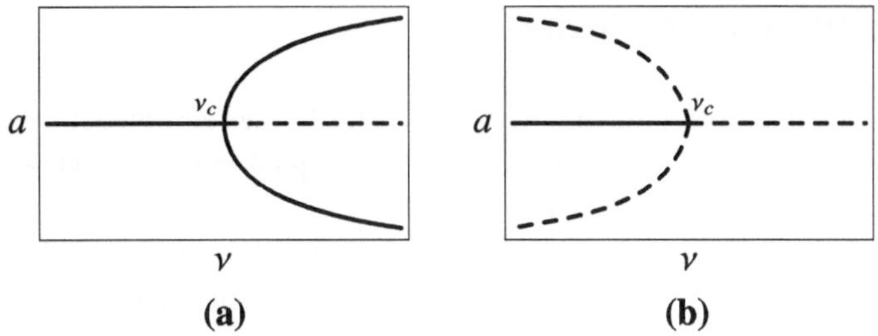

Fig. 2.2 Bifurcation diagram (pitchfork) for the symmetric beam, axially prestressed by a dead load: **a** $C_2 > 0$, **b** $C_2 < 0$

follows, where[6]:

$$C_2 := \frac{1}{2} k_3 \frac{\int_0^1 \phi_c^4 \mathrm{d}s}{\int_0^1 \phi_c'^2 \mathrm{d}s}. \tag{2.40}$$

The first order solution, therefore, is:

$$\begin{aligned} u &= a\phi_c(s), \\ \nu &= \nu_c + C_2 a^2, \end{aligned} \tag{2.41}$$

from which $u = \pm\phi_c(s)\sqrt{(\nu - \nu_c)/C_2}$. The relevant bifurcation diagram is shown in Fig. 2.2; it describes a *pitchfork bifurcation*. It is apparent that the bifurcated path lies on one side of the bifurcation point, according to the sign of C_2. When $C_2 > 0$ (Fig. 2.2a), the bifurcation is said *supercritical*, since buckled equilibria exist for $\nu > \nu_c$; when $C_2 < 0$ (Fig. 2.2b), the bifurcation is said *subcritical*, since buckled equilibria exist for $\nu < \nu_c$. We will detect later (see the Sect. 4.3), that the fundamental path loses stability at bifurcation, while bifurcated supercritical branches are stable, and bifurcated subcritical branches are unstable. Thus, again, an exchange of stability occurs at the bifurcation. Subcritical bifurcations are extremely dangerous, since they prevent the motion to be attracted by stable trivial equilibrium points, if an initial disturb is sufficiently large; in other words, they limit the *basin of attraction* of stable equilibria.

Remark 2.5 In symmetric systems, the change of shape of the buckling pattern, related to quadratic nonlinearities in non-symmetric systems, becomes a higher order effect. Therefore, Eq. 2.41 describes an asymptotically exact solution in which, along the bifurcated path, an homothetic increase of the beam deflection occurs.

[6] This result is consistent with that of the non-symmetric system (Eq. 2.32), when one takes into account that the current actual perturbation parameter is the square of the former one.

2.3.3 Imperfection Sensitivity

Transcritical and pitchfork bifurcations concern ideally 'perfect' models, namely straight and homogeneous beams, stressed by forces exactly aligned and applied at centerline. Since physical models are always affected by imperfections (e.g., small curvatures, small non-homogeneities, small eccentricities or inclinations of forces), we could ask if the previous results are *robust* under these disturbances, in the sense that their qualitative character persists or it is destroyed by perturbations.[7] We will see soon that, indeed, transcritical and pitchfork bifurcations disappear under small perturbations produced by imperfections, and therefore they are non-robust, i.e., they cannot be observed in the real world, e.g., in a laboratory.[8]

To investigate the effects of disturbances on static bifurcations, we consider the presence of a small transverse force αP_B applied at the tip of the beam, where $0 < \alpha \ll 1$ is the magnitude imperfection parameter, and $P_B = \mathrm{O}(1)$. The last boundary condition in Eqs. 2.16, therefore, becomes non-homogeneous, namely $-u'''(1) - 2\nu u'(1) = \alpha P_B$ and the previous perturbation procedure must be updated to account for the new term. First, we have to suitably rescale the small imperfection α. Since we rescaled the solution as $u \to \epsilon\hat{u}$, and we require the external force αP_B to be smaller than the elastic forces, u'''' and $k_1 u$ (these mainly produced by buckling and not by the force itself), we must rescale the imperfection as $\alpha \to \epsilon^m \hat{\alpha}$, with $m = 2, 3, \ldots$. The solution of the 'perfect' problem guides us in choosing m. Indeed, the main information about bifurcation is brought by quadratic nonlinearities, if the system is non-symmetric, and by cubic nonlinearities, if the system is symmetric. We observe that the imperfection significantly affects the solution *when it is competing with the main nonlinearities*, i.e., when $\alpha = \mathrm{O}(u^2)$, or $\alpha = \mathrm{O}(u^3)$, in the two classes of systems, respectively. Therefore, by taking $m = 2, 3$ in the two cases, the imperfection is shifted to the ϵ- or ϵ^2- perturbation equations, i.e., to Eqs. 2.20 or 2.21.

In the non-symmetric case, the last boundary condition in Eqs. 2.20 contains, on the right-hand side, the extra-term αP_B. Therefore, the compatibility condition Eq. 2.25 must be modified as follows:

$$-\int_0^1 \phi_c \left(2a\nu_1\phi_c'' + a^2 k_2 \phi_c^2\right) \mathrm{d}s + \phi_c(1)[2a\nu_1\phi_c'(1) + \alpha P_B] = 0, \tag{2.42}$$

and Eq. 2.26 becomes:

$$\nu_1 = C_1 a + \frac{\alpha}{a} C_\alpha, \tag{2.43}$$

where:

$$C_\alpha := -\frac{P_B}{2}\frac{\phi_c(1)}{\int_0^1 \phi_c'^2 \mathrm{d}s}. \tag{2.44}$$

---

[7] In Stability Theory, when a bifurcation persists under variations of parameters, it is called *structurally stable*; therefore robustness is synonymous of structural stability.

[8] In contrast, it can be proved, that a *limit* (or fold) *point* is robust.

In the symmetric case, the extra-term αP_B instead appears in the last boundary condition of Eqs. 2.21. Hence, compatibility Eq. 2.38 is updated to:

$$-\int_0^1 \phi_c \left(2a\nu_1\phi_c'' + a^3k_3\phi_c^3\right) \mathrm{d}s + \phi_c(1)[2a\nu_1\phi_c'(1) + \alpha P_B] = 0, \tag{2.45}$$

from which Eq. 2.39 is modified into:

$$\nu_1 = C_2a^2 + \frac{\alpha}{a}C_\alpha. \tag{2.46}$$

If both analyses are truncated at the first order, the following load-amplitude-imperfection laws are found:

$$\nu = \begin{cases} \nu_c + C_1a + \frac{\alpha}{a}C_\alpha, & \text{if } k_2 \neq 0, \\ \nu_c + C_2a^2 + \frac{\alpha}{a}C_\alpha, & \text{if } k_2 = 0. \end{cases} \tag{2.47}$$

Equation 2.47 describe surfaces in the (ν, a, α)-space. They can be represented by contour lines $\alpha = \text{const}$ in the (ν, a)-plane (Fig. 2.3). It can be seen that, irrespective of the magnitude of the imperfection, the branch points are destroyed. Indeed, there are neither fundamental nor bifurcated paths, as they existed for the perfect system, but only *nontrivial* paths. When

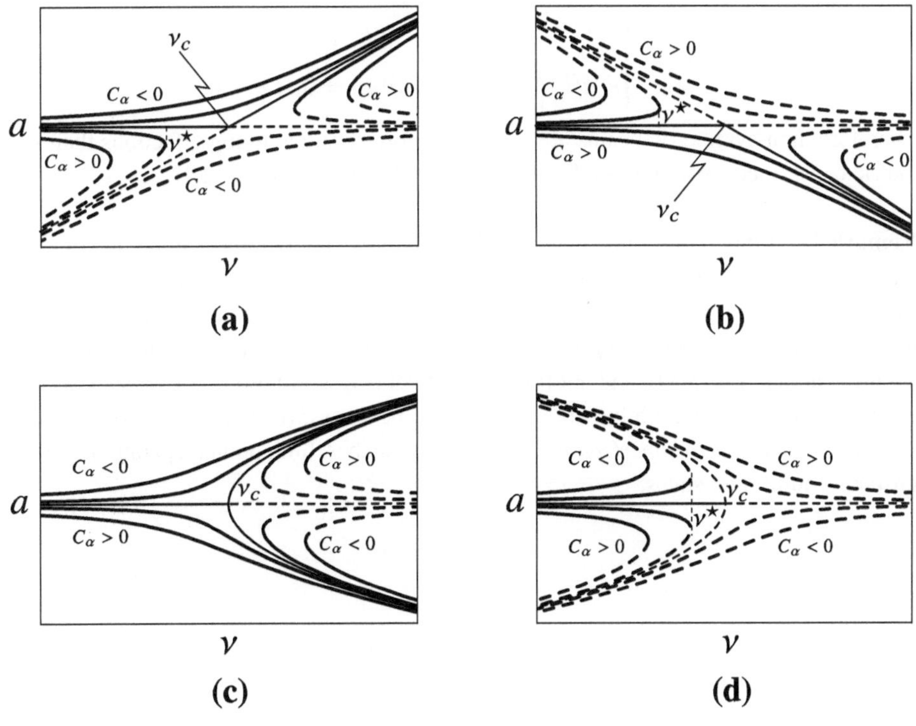

Fig. 2.3 Effect of imperfections α on: **a, b** transcritical, **c, d** pitchfork bifurcations; $\nu^\star$: limit loads

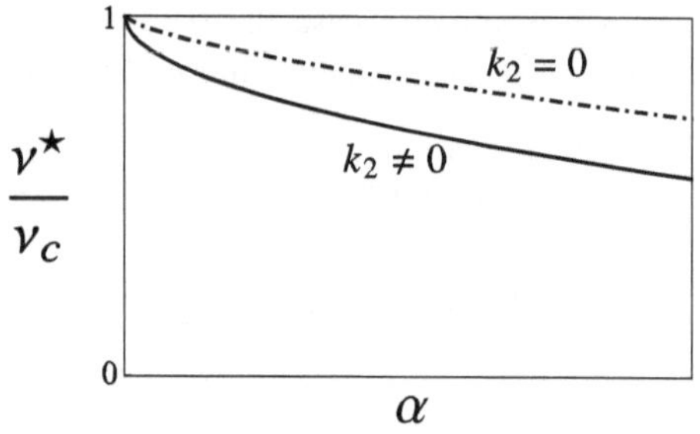

Fig. 2.4 Erosion of the stability due to imperfection α for non-symmetric (continuous line) and symmetric (dash-dotted line) systems undergoing sub-critical bifurcation ($C_2 < 0$)

the system is non-symmetric (Fig. 2.3a, b), a new form of instability is displayed, namely the *fold point*, better known in the engineering community as *limit point*.[9] Here, a local maximum exists for the prestress load, said the *limit load* $\nu = \nu^\star(\alpha)$, which depends on α. Following the equilibrium path, once the limit point is reached, the solution *jumps* elsewhere if the prestress is slightly increased.

Imperfections produce a detrimental effect on the mechanical behavior of the beam (sometimes called *erosion* of stability [2]), since they lower the threshold at which the equilibrium path loses stability, namely from the bifurcation value ν_c of the perfect system to the limit value $\nu^\star(\alpha) < \nu_c$ of the imperfect (sensitive) system. From Eq. 2.47, it follows that:

$$\nu^\star - \nu_c = \begin{cases} -2\alpha^{1/2}\sqrt{C_1 C_\alpha}, & \text{if } k_2 \neq 0,\ C_1 C_\alpha > 0, \\ -\alpha^{2/3}\sqrt[3]{-\frac{27}{4}C_2 C_\alpha^2}, & \text{if } k_2 = 0,\ C_2 < 0. \end{cases} \tag{2.48}$$

Therefore, non-symmetric beams are generally more sensitive to imperfections than symmetric ones, since erosion depends on $\alpha^{1/2}$, instead of $\alpha^{2/3}$ (see Fig. 2.4).

Remark 2.6 Limit points of non-symmetric systems exist only for a suitable sign of imperfections, but, since imperfections are unpredictable, the worse case must be taken into account for design purposes. In contrast, when the system is symmetric (and restricting to the *natural curves* emanating from the $\nu = 0$ axis), limit point exists only for subcritical bifurcations (Fig. 2.3d), and not for supercritical (Fig. 2.3c). For these reasons, systems experiencing supercritical bifurcations are said to be *insensitive to imperfections*; all other systems are *sensitive to imperfections*. Along any branches, limit points separate stable from unstable equilibrium states.

[9] Also said *saddle-node* in the nonlinear dynamics community.

2.4 Exercises

The perturbation methods illustrated in this chapter are applied to some case studies. Goals of the analysis are: (i) to evaluate the response of the beam under uniform transverse load, and (ii) to build-up the bifurcation diagram of the compressed beam. A preparatory exercise, concerning the solution of non-homogeneous linear differential equations, useful for later developments, is premised.

Exercise 2.1 (*Non-homogeneous linear differential equation*) The following linear, fourth order, non-homogeneous ordinary differential equation, and relevant boundary conditions, is given:

$$\chi''''(s) + \kappa\chi(s) = p(s), \tag{2.49a}$$

$$\chi(0) = 0, \qquad \chi'(0) = 0, \tag{2.49b}$$

$$\chi''(1) = 0, \qquad -\chi'''(1) = P, \tag{2.49c}$$

in which $\kappa > 0$. (a) Derive the particular solution using the method of variations of constants. (b) Evaluate and plot the solution for $p(s) \equiv \bar{p} = 0.1$, $P = 0.1$ and $\kappa = 1$.

(a) First, the general solution of the associated homogeneous problem is evaluated, using the test solution $\chi(s) = e^{\lambda s}$. As a consequence, the characteristic equation is obtained:

$$\lambda^4 + \kappa = 0, \tag{2.50}$$

which has solutions[10]:

$$\lambda_{1,2} = \pm i\beta, \qquad \lambda_{3,4} = \pm\beta, \tag{2.51}$$

where $\beta := \tilde{\beta}\sqrt{i}$ is a *complex* number, with $\tilde{\beta} := \sqrt[4]{\kappa}$. The general solution is a linear combination of the obtained test solutions, namely[11]:

$$\chi(s) = a_1\cos(\beta s) + a_2\sin(\beta s) + a_3\cosh(\beta s) + a_4\sinh(\beta s) =: \sum_{j=1}^{4} a_j x_j(s), \tag{2.52}$$

[10] Indeed, $\lambda^2 = \pm i\sqrt{\kappa}$. By taking the negative sign, $\lambda_{1,2} = \pm\sqrt{-1}\sqrt{i}\sqrt[4]{\kappa}$ follows; by taking the positive sign, $\lambda_{3,4} = \pm\sqrt{i}\sqrt[4]{\kappa}$ is obtained.

[11] An alternative way to obtain the general solution of Eqs. 2.49, made homogeneous, consists in using real quantities, as follows. By defining a new parameter ζ, such that $\kappa = 4\zeta^4$, the characteristic Eq. 2.50 reads $\lambda^4 + 4\zeta^4 = 0$, whose solutions are $\lambda_{1,2,3,4} = \pm(1 \pm i)\zeta$. Hence, the general solution assumes the form:

$$\chi(s) = a_1 e^{\zeta s}\cos(\zeta s) + a_2 e^{\zeta s}\sin(\zeta s) + a_3 e^{-\zeta s}\cos(\zeta s) + a_4 e^{-\zeta s}\sin(\zeta s).$$

Although this equation is real, it leads to a more complicated Wronskian matrix, which generally can make more difficult obtaining of the particular solutions. Therefore, it is preferred here, to use complex quantities, and to resort to the real ones only at the end of the procedure.

where $x_1(s) := \cos(\beta s)$, $x_2(s) := \sin(\beta s)$, $x_3(s) := \cosh(\beta s)$, $x_4(s) := \sinh(\beta s)$ and a_j ($j = 1, \dots 4$) are arbitrary constants.

In order to solve the non-homogeneous Eqs. 2.49, a solution of the following form is sought[12]:

$$\chi(s) = \sum_{j=1}^{4} \tilde{a}_j(s) x_j(s), \tag{2.53}$$

where the s-depending coefficients $\tilde{a}_j(s)$ have to be evaluated; in particular, they are chosen in such a way the following equations hold:

$$\sum_{j=1}^{4} \tilde{a}'_j(s) x_j(s) = 0, \tag{2.54a}$$

$$\sum_{j=1}^{4} \tilde{a}'_j(s) x'_j(s) = 0, \tag{2.54b}$$

$$\sum_{j=1}^{4} \tilde{a}'_j(s) x''_j(s) = 0. \tag{2.54c}$$

As a consequence of Eqs. 2.54, differentiation of Eq. 2.53 gives:

$$\chi'(s) = \sum_{j=1}^{4} \tilde{a}_j(s) x'_j(s), \tag{2.55a}$$

$$\chi''(s) = \sum_{j=1}^{4} \tilde{a}_j(s) x''_j(s), \tag{2.55b}$$

$$\chi'''(s) = \sum_{j=1}^{4} \tilde{a}_j(s) x'''_j(s), \tag{2.55c}$$

$$\chi''''(s) = \sum_{j=1}^{4} [\tilde{a}'_j(s) x'''_j(s) + \tilde{a}_j(s) x''''_j(s)]. \tag{2.55d}$$

After substituting Eqs. 2.55d and 2.53 in Eq. 2.49a, an by taking into account that $x_j(s)$ are solutions of the homogeneous equation, the following condition is obtained:

$$\sum_{j=1}^{4} \tilde{a}'_j(s) x'''_j(s) = p(s). \tag{2.56}$$

---

[12] The method is known as the *Lagrange's method of variation of the constants.*

Equations 2.54a–c, 2.56 form an algebraic non-singular non-homogeneous system in the unknowns $\tilde{a}'_j(s)$, which can be put in matrix form:

$$\mathbf{W}(s)\tilde{\mathbf{a}}'(s) = \mathbf{p}(s), \tag{2.57}$$

where:

$$\mathbf{W}(s) = \begin{pmatrix} x_1(s) & x_2(s) & x_3(s) & x_4(s) \\ x_1'(s) & x_2'(s) & x_3'(s) & x_4'(s) \\ x_1''(s) & x_2''(s) & x_3''(s) & x_4''(s) \\ x_1'''(s) & x_2'''(s) & x_3'''(s) & x_4'''(s) \end{pmatrix} \tag{2.58}$$

is called the *Wronskian matrix*, $\mathbf{p}(s) := (0\,0\,0\,p(s))^T$ and $\tilde{\mathbf{a}}'(s) := (\tilde{a}_1'(s)\,\tilde{a}_2'(s)\,\tilde{a}_3'(s)\,\tilde{a}_4'(s))^T$.

Solution to system of Eqs. 2.57 can be found using the Cramer rule:

$$\tilde{a}'_j(s) = \frac{\det \mathbf{W}_j(s)}{\det \mathbf{W}(s)}, \tag{2.59}$$

where $W_j(s)$ is obtained by substituting the jth column of $W(s)$ by $p(s)$. In order to evaluate $\tilde{a}_j(s)$, integration of Eq. 2.59 is performed, leading to:

$$\tilde{a}_j(s) = \int \frac{\det \mathbf{W}_j(s)}{\det \mathbf{W}(s)} \mathrm{d}s. \tag{2.60}$$

(b) For the simple case $p(s) \equiv \bar{p} = \text{const}$, the following expressions for the coefficients are found:

$$\begin{aligned} \tilde{a}_1(s) &= -\frac{\bar{p}}{2\beta^4}\cos(\beta s), & \tilde{a}_2(s) &= -\frac{\bar{p}}{2\beta^4}\sin(\beta s), \\ \tilde{a}_3(s) &= -\frac{\bar{p}}{2\beta^4}\cosh(\beta s), & \tilde{a}_4(s) &= \frac{\bar{p}}{2\beta^4}\sinh(\beta s), \end{aligned} \tag{2.61}$$

so that, by using Eq. 2.53, the particular solution reads[13]:

$$\chi(s) = -\frac{\bar{p}}{\beta^4}. \tag{2.62}$$

The general solution to the ordinary differential equation is the sum of the solutions in Eqs. 2.52 and 2.62, i.e.:

$$\chi(s) = a_1\cos(\beta s) + a_2\sin(\beta s) + a_3\cosh(\beta s) + a_4\sinh(\beta s) - \frac{\bar{p}}{\beta^4}. \tag{2.63}$$

[13] As an alternative method, when the load is uniform, the particular solution can be evaluated using the uniform test function $\chi(s) \equiv \bar{\chi}$ in Eq. 2.49a, thus obtaining $\bar{\chi} = \frac{\bar{p}}{\kappa}$. Similarly, when the load is harmonic, e.g., $p(s) = \bar{p}\cos(\Lambda s)$, the test function $\chi(s) \equiv \bar{\chi}\cos(\Lambda s)$ can be used, leading to the algebraic equation $\Lambda^4\bar{\chi} + \kappa\bar{\chi} = \bar{p}$, which provides $\bar{\chi} = \frac{\bar{p}}{\kappa}\frac{1}{1+\frac{\Lambda^4}{\kappa}}$. For more general known terms, however, as it happens in solving higher order perturbation equations, the method illustrated here is needed.

When the boundary conditions Eqs. 2.49b, c are enforced, the constants a_j are computed as:

$$\begin{aligned}
a_1 =& \frac{\bar{p} + \bar{p}\cos\beta\cosh\beta - P\beta\sin\beta - (P\beta + \bar{p}\sin\beta)\sinh\beta}{2\beta^4(1+\cos\beta\cosh\beta)},\\
a_2 =& \frac{(P\beta + \bar{p}\sin\beta)\cosh\beta + (P\beta + \bar{p}\sinh\beta)\cos\beta}{2\beta^4(1+\cos\beta\cosh\beta)},\\
a_3 =& \frac{\bar{p} + \bar{p}\cos\beta\cosh\beta + P\beta\sin\beta + (P\beta + \bar{p}\sin\beta)\sinh\beta}{2\beta^4(1+\cos\beta\cosh\beta)},\\
a_4 =& -\frac{(P\beta + \bar{p}\sin\beta)\cosh\beta + (P\beta + \bar{p}\sinh\beta)\cos\beta}{2\beta^4(1+\cos\beta\cosh\beta)}.
\end{aligned} \tag{2.64}$$

Finally, to get a real expression for the solution of the boundary value problem, since $\beta = \tilde{\beta}\sqrt{i}$, the following trigonometric identities must be used, with $\gamma := \tilde{\beta}$ or $\gamma := \tilde{\beta}s$:

$$\begin{aligned}
\cos(\gamma\sqrt{i}) &= \cos\left(\frac{\gamma}{\sqrt{2}}\right)\cosh\left(\frac{\gamma}{\sqrt{2}}\right) - i\sin\left(\frac{\gamma}{\sqrt{2}}\right)\sinh\left(\frac{\gamma}{\sqrt{2}}\right),\\
\sin(\gamma\sqrt{i}) &= \cosh\left(\frac{\gamma}{\sqrt{2}}\right)\sin\left(\frac{\gamma}{\sqrt{2}}\right) + i\cos\left(\frac{\gamma}{\sqrt{2}}\right)\sinh\left(\frac{\gamma}{\sqrt{2}}\right),\\
\cosh(\gamma\sqrt{i}) &= \cos\left(\frac{\gamma}{\sqrt{2}}\right)\cosh\left(\frac{\gamma}{\sqrt{2}}\right) + i\sin\left(\frac{\gamma}{\sqrt{2}}\right)\sinh\left(\frac{\gamma}{\sqrt{2}}\right),\\
\sinh(\gamma\sqrt{i}) &= \cos\left(\frac{\gamma}{\sqrt{2}}\right)\sinh\left(\frac{\gamma}{\sqrt{2}}\right) + i\cosh\left(\frac{\gamma}{\sqrt{2}}\right)\sin\left(\frac{\gamma}{\sqrt{2}}\right).
\end{aligned} \tag{2.65}$$

A plot of the solution is shown in Fig. 2.5 for $\kappa = 1$, $\bar{p} = 0.1$, $P = 0.1$. □

Exercise 2.2 (*Static response of a beam on elastic soil and uniform load*) Determine the displacement field of a beam on non-symmetric elastic soil, under time-independent and uniform load $p(s) \equiv \bar{p}$; use an asymptotic expansion up to the order ϵ^2.

The displacement of the beam is expressed by Eq. 2.10, where $\chi_m(s)$ $(m = 0, 1, 2)$ are solutions to Eqs. 2.9, with $p(s) \equiv \bar{p}$ and $P_B \equiv 0$. Making use of the results of Exercise 2.1, we can evaluate the ϵ^0 order solution:

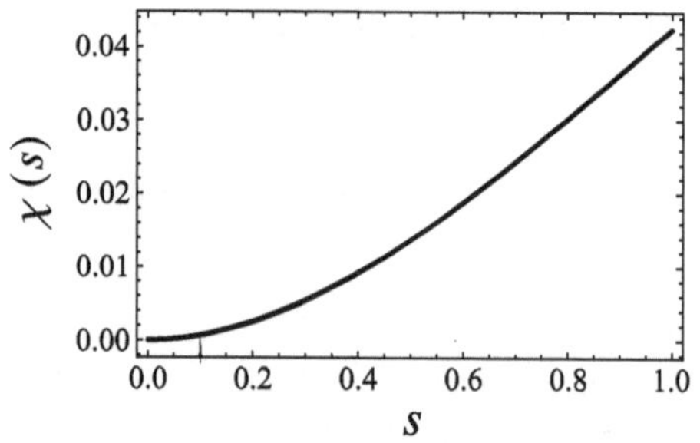

Fig. 2.5 Solution to the boundary value problem in Eq. 2.49, when $\kappa = 1$, $\bar{p} = 0.1$, $P = 0.1$

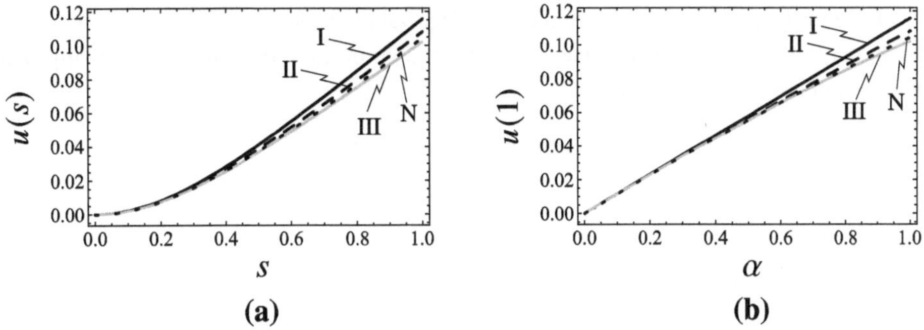

Fig. 2.6 Solution to the boundary value problem in Eqs. 2.9, when $k_1 = 1$, $k_2 = 10$, $k_3 = 100$, $\bar{p} = 1$: **a** displacement of the beam when $\alpha = 1$; **b** displacement of the tip of the beam *versus* the load multiplier α. (I) order ϵ^0 solution (continuous line); (II) order ϵ^1 solution (dashed line); (III) order ϵ^2 solution (gray line); (N) numerical solution of the nonlinear problem in the Eq. 2.1 (dotted line)

$$\chi_0(s) = a_1 \cos(\beta s) + a_2 \sin(\beta s) + a_3 \cosh(\beta s) + a_4 \sinh(\beta s) - \frac{\bar{p}}{\beta^4}, \tag{2.66}$$

where $\beta = \sqrt{i}\,\sqrt[4]{k_1}$ and the constants a_j are given by Eqs. 2.64 with $P \equiv 0$.

As a second step, the ϵ^1 order solution $\chi_1(s)$ is evaluated by considering both the complementary and particular solutions of system, Eqs. 2.9, with $m = 1$. In the same way, the third step is performed evaluating the ϵ^2 order solution $\chi_2(s)$ from Eq. 2.9, with $m = 2$. At each step, the boundary conditions are used to evaluate the coefficients of the relevant complementary part of the solution.[14]

The displacement $u(s)$ of the beam is shown in Fig. 2.6a for the following values of the parameters: $k_1 = 1, k_2 = 10, k_3 = 100, \bar{p} = 1$ and the load multiplier $\alpha = 1$, considering truncated expansions, (i) at the order ϵ^0 (continuous line), (ii) at the order ϵ^1 (dashed line), and (iii) at the order ϵ^2 (gray line). In Fig. 2.6b, the displacement of the tip of the beam, $u(1)$, is shown as a function of the load multiplier. The asymptotic solutions of Fig. 2.6 are compared with a numerical solution (N) of the nonlinear problem in the Eq. 2.1 (dotted line). □

Exercise 2.3 (*Buckling and postbuckling of a compressed beam on elastic soil*) Consider a beam on non-symmetric Winkler soil, with $k_1 = 1$, $k_2 = 10$ and $k_3 = 100$, prestressed by a dead load ν. (a) Evaluate the critical load ν_c and the critical mode $\phi_c(s)$. (b) Compute the displacement field $u(s)$ for a load higher than the critical one. (c) Plot a bifurcation diagram, depicting the displacement of the tip of the beam as a function of ν.

---

[14] It is worth noticing that, as an alternative strategy of resolution, pure numerical solutions of Eqs. 2.9 for $\chi_j(s)$ can be sought using classical algorithms for boundary value problems. Finally, Eq. 2.10 can be used to describe the dependence of the reconstituted solution $\chi(s)$ on the load parameter α, which is made explicit.

(a) To evaluate the critical load, the boundary value problem Eqs. 2.19 must be solved. To do that, the test solution $u_0(s) = e^{\lambda s}$ is substituted in Eq. 2.19a, and the following characteristic equation is obtained:

$$\lambda^4 + 2\nu_0\lambda^2 + k_1 = 0. \tag{2.67}$$

This admits the roots:

$$\lambda_{1,2} = \pm\sqrt{-\nu_0 + \sqrt{\nu_0^2 - k_1}}, \quad \lambda_{3,4} = \pm\sqrt{-\nu_0 - \sqrt{\nu_0^2 - k_1}}. \tag{2.68}$$

Under the hypothesis $\nu_0^2 > k_1$ (to be verified later), the characteristic exponents λ_i are two couples of pure imaginary numbers, namely: $\lambda_{1,2} = \pm i\beta_1$, $\lambda_{3,4} = \pm i\beta_2$, where $\beta_1 := \sqrt{\nu_0 + \sqrt{\nu_0^2 - k_1}}$ and $\beta_2 := \sqrt{\nu_0 - \sqrt{\nu_0^2 - k_1}}$. Therefore, using the Euler formula, the general solution of Eq. 2.19a is[15]:

$$u_0(s) = a_1 \cos(\beta_1 s) + a_2 \sin(\beta_1 s) + a_3 \cos(\beta_2 s) + a_4 \sin(\beta_2 s). \tag{2.69}$$

When it is used in the homogeneous boundary conditions, Eqs. 2.19b, c, these latter lead to a homogeneous algebraic problem in the unknowns a_j, of type $\mathbf{Aa} = \mathbf{0}$, where $\mathbf{a} = (a_1\ a_2\ a_3\ a_4)^T$. The critical loads are the values of ν_0 which make the matrix $\mathbf{A}$ singular; Fig. 2.7 shows a plot of $\det \mathbf{A}$ *versus* ν_0, when $k_1 = 1$, whose intersection with the ν_0 axis are the roots sought for. The lower critical load is found to be $\nu_c = 11.1905$ (for which the starting hypothesis $\nu_0^2 > k_1$ is satisfied); the corresponding critical mode $\phi_c(s)$, where the normalization $\phi_c(1) = 1$ has been used, is shown in Fig. 2.7b by a continuous line.

(b) The first order contribution to the bifurcated path is given by Eq. 2.26, where the constant assume the value $C_1 = 1.4712$. To obtain the second order correction $u_1(s)$, the singular problem in Eq. 2.29 must be solved (for which solvability has already been satisfied), to get the expression of $\chi_1(s)$. To do that, first, the particular solution of Eq. 2.29a is sought, making use of the results of the Exercise 2.1, where the Wronskian matrix is built-up through Eq. 2.69. Then, this particular solution is summed to the general solution Eq. 2.69, in order to satisfy the boundary conditions in Eqs. 2.29b, c. However, due to the singularity of the problem, the boundary conditions are not all independent, so that one of them must be

[15] The hypothesis $\nu_0^2 > k_1$ is mostly verified for higher critical conditions (i.e., for larger ν_0) than the first one. For the lower critical conditions, if the hypothesis is not verified, then β_1 and β_2 become complex conjugate numbers, namely $\beta_{1,2} = \beta_R(\nu_0) \pm i\beta_I(\nu_0)$ and Eq. 2.69 for the eigenvector should be substituted by the following real expression:

$$\begin{aligned} u_0(s) =& a_1 \cos(\beta_R s)\sinh(\beta_I s) + a_2 \sin(\beta_R s)\sinh(\beta_I s) \\ &+ a_3 \cos(\beta_R s)\cosh(\beta_I s) + a_4 \sin(\beta_R s)\cosh(\beta_I s). \end{aligned}$$

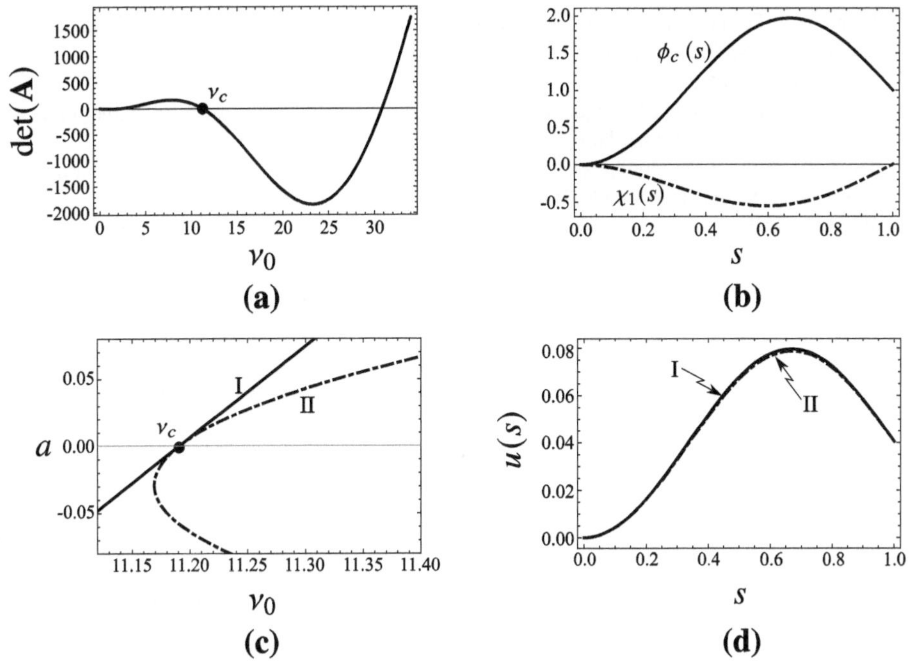

Fig. 2.7 Solution of the boundary value problem in Eqs. 2.16, when $k_1 = 1$, $k_2 = 10$, $k_3 = 100$: **a** determinant of the matrix **A** as a function of ν_0; **b** critical mode ϕ_c and passive function χ_1; **c** bifurcation diagram; **d** postcritical displacement of the beam when $\nu = 11.25$; (I) order ϵ^0 solution (continuous line:); (II) order ϵ^1 solution (dash-dotted line)

replaced by the normalization condition $\chi_1(1) = 0$. The resulting function $\chi_1(s)$ is shown in Fig. 2.7b by a dash-dotted line.

(c) The second order contribution to the bifurcated path is given by Eq. 2.32, in which $C_2 = 26.81$. The bifurcation diagram is shown in Fig. 2.7c, where the linear (I) and quadratic (II) approximations are highlighted. When a supercritical load is considered, e.g., $\nu = 11.25$, the linear (I) and linear plus quadratic (II) displacements of the beam are shown in Fig. 2.7d, corresponding to the bifurcated paths of Fig. 2.7c. □

References

1. Carr, J.: Applications of Centre Manifold Theory. Springer, New York (1981)
2. Gioncu, V.: General theory of coupled instabilities. Thin Walled Struct. **19**(2–4), 81–127 (1994)
3. Guckenheimer, J., Holmes, P.: Nonlinear Oscillations, Dynamical Systems, and Bifurcations of Vector Fields. Springer, New York (1983)

4. Luongo, A., Ferretti, M., Di Nino, S.: Stability and Bifurcation of Structures: Statical and Dynamical Systems. Springer, Cham (2023)
5. Nayfeh, A.H.: Perturbation Methods. Wiley, New York (1973)
6. Pignataro, M., Rizzi, N., Luongo, A.: Stability, Bifurcation and Postcritical Behaviour of Elastic Structures. Elsevier, Amsterdam (1990)
7. Troger, H., Steindl, A.: Nonlinear Stability and Bifurcation Theory: An Introduction for Engineers and Applied Scientists. Springer, Wien (1991)
8. Wiggins, S.: Introduction to Applied Nonlinear Dynamical Systems and Chaos. Springer, New York (1990)

Non-autonomous Dynamical Systems 3

3.1 Introduction

Elastic structures subjected to external loads, varying in time with harmonic laws, are non-autonomous systems prone to several resonance phenomena. Two main classes of problems must be distinguished: (a) *external resonances*, occurring when the load appears in the right hand member of the equation as a know term; (b) *parametric resonances*, manifesting when the load appears, as a geometric effect, in the left member of the equation (as in the Euler beam), thus making the coefficients time-varying.

The external resonances occur when the forcing frequency is in particular ratios with one of the natural frequencies. The most known (and important) one is the *primary resonance*, in which such a ratio is equal to 1. Differently from the linear theory, which predicts an infinite response (in absence of damping), the nonlinear theory describes large, but not infinite (and also multivalued) responses, whose evaluation is of great interest. In addition, *sub-harmonic* and *super-harmonic resonances*, which do not exist in the linear theory, are also encountered, occurring when the natural-to-excitation frequency ratio is a rational number, respectively smaller or larger than 1. However, while the primary resonance manifests even for small amplitudes of the load (*soft excitation*), the sub- and super-harmonic resonances require loads of large amplitude (*hard excitation*).

The parametric excitation is a phenomenon suffered also by linear systems. It manifests in certain *tongues of instability* existing in the amplitude-frequency parameter plane. Inside these regions, which emanate (in absence of damping) from the frequency axis, the trivial equilibrium position is unstable. The nonlinear theory permits to evaluate the amplitude of the limit cycle, bifurcating from the trivial solution at the boundaries of the tongues.

All these phenomena are here analyzed by using the *Multiple Scale Method*, which is an asymptotic method that supplies bifurcation equations in the active variables, namely in the amplitude and phase of the natural mode involved in the resonance. Here it is illustrated

A. Luongo et al., *Perturbation Methods and Nonlinear Phenomena*,
Synthesis Lectures on Engineering, Science, and Technology,
https://doi.org/10.1007/978-3-031-49397-3_3

referring to the sample system (the beam on Winkler soil), by limiting the discussion to the simplest cases, in which no internal resonances are involved, all leading to two-dimensional bifurcation equations.

3.2 Primary External Resonance

In this section we will focus the attention on a linear beam, resting on nonlinear elastic soil, externally excited by harmonic varying *transverse* forces, pulsating at a frequency Ω close to one of the natural frequencies of the beam, e.g., ω_r. This problem is called of *primary resonance*.[1]

3.2.1 Non-symmetric Systems

Let us consider the motion of an unprestressed beam ($\mu = \nu_s = \nu_d = 0$) on elastic soil, under transverse distributed and point loads, varying in time with harmonic law, $p(s, t) = q(s)\cos(\Omega t)$, $P_B(t) = Q_B \cos(\Omega t)$. The equations of motion, Eq. 1.7 specialize as:

$$\ddot{u} + \xi \dot{u} + u'''' + k_1 u + k_2 u^2 + k_3 u^3 + \cdots = \frac{1}{2}\alpha q(s) e^{i\Omega t} + \text{c.c.}, \tag{3.1a}$$

$$u(0, t) = 0, \quad u'(0, t) = 0, \tag{3.1b}$$

$$u''(1, t) = 0, \quad -u'''(1, t) = \frac{1}{2}\alpha Q_B e^{i\Omega t} + \text{c.c.}. \tag{3.1c}$$

where c.c. stands for the complex conjugate of preceding terms.

Natural Frequencies and Modes

When $\alpha = 0, \xi = 0$ and nonlinearities are ignored, the previous equations govern the (small) free oscillations of the undamped beam. They admit the solution $u(s, t) = \phi_k(s)e^{i\omega_k t}$, with (ω_k, ϕ_k) the kth ($k = 1, 2, \ldots$) natural frequency and normal mode, respectively. These latter are an *eigenpair* (i.e., an eigenvalue with associated eigenvector) of the following boundary value problem:

$$\phi_k'''' + (k_1 - \omega_k^2)\phi_k = 0, \tag{3.2a}$$

$$\phi_k(0) = 0, \quad \phi_k'(0) = 0, \tag{3.2b}$$

$$\phi_k''(1) = 0, \quad -\phi_k'''(1) = 0. \tag{3.2c}$$

[1] A discrete version of the problem is dealt with in the Sect. B.4.1 of the Appendix B.

Since this problem is self-adjoint (being $\mu = 0$), then right and left eigenvectors coincide, i.e., $\psi_k \equiv \phi_k$ (see the Appendix A).

Rescaling

Let us assume that the excitation frequency Ω is close to the rth natural frequency ω_r, i.e., let it be:

$$\Omega = \omega_r + \sigma_r, \tag{3.3}$$

where σ_r is a small *detuning*, so that a primary external resonance occurs. As it is well-known, the response of the beam diverges at infinity in the linear theory when $\sigma_r \to 0$. However, nonlinearities prevent this phenomenon, by limiting the amplitude of the response, in a way we want to investigate. Since we consider that nonlinearities are small (in order a perturbation approach is viable), we have to assume that the amplitude α of the excitation is also small, of the same order of nonlinearities.

To tackle the problem, we will use the *Multiple Scale Method* (MSM, [3]). Differently from the strained parameter method, in which a parameter is expanded, *here parameters are not expanded, but simply rescaled*. The reasons of this procedure will be commented ahead (see Remark 3.2). According to the method, the displacement $u(s,t)$, as well as the small coefficients appearing in the equation, are rescaled, namely: the excitation amplitude α, the detuning σ_r, the damping ξ. We introduce the following rescaling:

$$u \to \epsilon \hat{u}, \quad \alpha \to \epsilon^3 \hat{\alpha}, \quad \sigma_r \to \epsilon^2 \hat{\sigma}_r, \quad \xi \to \epsilon^2 \hat{\xi}. \tag{3.4}$$

It appears that not all the quantities have been equally rescaled, but, in contrast, different powers of ϵ have been used for each quantity, aimed to 'properly' place the various terms in different perturbation equations. These choices usually represent the main difficulty of the method, and they will appear clear only a posteriori.[2] So far, we limit ourselves to observe that: (a) ordering the excitation as the cube of the response means that we are looking for responses remarkably larger than the excitation, as it is expected to occur at resonance; (b) ordering the detuning as $\sigma_r = \mathrm{O}\left(\epsilon^2\right)$ means to consider quite 'narrow' resonances, i.e., to assume that quasi-resonance (at $\sigma \neq 0$) manifests itself in a small neighborhood of the 'perfect' resonance (occurring at $\sigma_r = 0$). Using Eqs. 3.4 and dividing by ϵ transforms Eqs. 3.1 into:

---

[2] This procedure could leave the novice reader astonished, but it encompasses the true essence of the perturbation method, that often requires to discover the 'right way' to (fast) reach the goal. Usually one has to perform several attempts, before finding the correct ordering and/or expansion. Of course, more 'automatic' versions of the method exist, in which one introduces several additional parameters, and successively vanishes the useless or malicious ones. However, this is a very inefficient application of the algorithm, that, therefore, we will avoid. Using the perturbation method, requires the analyst to heuristically approach the problem, in the sense that, if the method is internally consistent (e.g., not interrupting itself at a certain order, or leading to equations impossible to solve), then the solution is supposed to be correct. Confidence in results, usually, is rooted in numerical validations.

$$\ddot{u} + u'''' + k_1 u + \epsilon k_2 u^2 + \epsilon^2 (\xi \dot{u} + k_3 u^3) + \cdots = \frac{1}{2}\epsilon^2 \alpha q(s) e^{i(\omega_r + \epsilon^2 \sigma_r)t} + \text{c.c.}, \tag{3.5a}$$

$$u(0, t) = 0, \quad u'(0, t) = 0, \tag{3.5b}$$

$$u''(1, t) = 0, \quad -u'''(1, t) = \frac{1}{2}\epsilon^2 \alpha Q_B e^{i(\omega_r + \epsilon^2 \sigma_r)t} + \text{c.c.}. \tag{3.5c}$$

Multiple Scale Method

According to the MSM basic idea, the solution $u(s, t)$ of Eqs. 3.5 is assumed as a periodic signal, which is slowly modulated on N slower independent time-scales[3]:

$$t_0 := t, \quad t_1 := \epsilon t, \quad t_2 := \epsilon^2 t, \quad \ldots . \tag{3.6}$$

Accordingly, $u(s, t) = u(s, t_0(t), t_1(t), t_2(t), \ldots, t_N(t))$, so that, by the chain rule, it follows that:

$$\partial_t u = \left(\partial_0 + \epsilon \partial_1 + \epsilon^2 \partial_2 + \cdots\right) u, \tag{3.7a}$$

$$\partial_t^2 u = \partial_t(\partial_t u) = \left(\partial_0 + \epsilon \partial_1 + \epsilon^2 \partial_2 + \cdots\right)^2 u, \tag{3.7b}$$

where, to short the notation, $\partial_k := \partial_{t_k}$ has been used. Hence:

$$\partial_t = \partial_0 + \epsilon \partial_1 + \epsilon^2 \partial_2 + \cdots , \tag{3.8a}$$

$$\partial_t^2 = \partial_0^2 + 2\epsilon \partial_0 \partial_1 + \epsilon^2 \left(\partial_1^2 + 2\partial_0 \partial_2\right) + \cdots , \tag{3.8b}$$

are *formal expansion for the first and second time-differential operators*. On the other hand, the series expansion for the variable, as usual, reads:

$$u(s, t_k) = u_0(s, t_k) + \epsilon u_1(s, t_k) + \epsilon^2 u_2(s, t_k) + \cdots , \tag{3.9}$$

with $k = 1, \ldots, N$. By substituting the series of Eqs. 3.8, 3.9 in the rescaled Eqs. 3.5, and separately equating to zero terms with the same power of ϵ, the following perturbation equations are finally derived (up to order ϵ^2):

[3] An example in nature is offered by the law in which the temperature of a site varies with time. On a short-time monitoring, it appears periodic with daily period, but after longer observation it appears modulated on a yearly-scale, and this, in turn, modulated on a century scale.

Order ϵ^0:

$$\partial_0^2 u_0 + u_0'''' + k_1 u_0 = 0, \tag{3.10a}$$

$$u_0(0, t_k) = 0, \quad u_0'(0, t_k) = 0, \tag{3.10b}$$

$$u_0''(1, t_k) = 0, \quad -u_0'''(1, t_k) = 0. \tag{3.10c}$$

Order ϵ^1:

$$\partial_0^2 u_1 + u_1'''' + k_1 u_1 = -2\partial_0\partial_1 u_0 - k_2 u_0^2, \tag{3.11a}$$

$$u_1(0, t_k) = 0, \quad u_1'(0, t_k) = 0, \tag{3.11b}$$

$$u_1''(1, t_k) = 0, \quad -u_1'''(1, t_k) = 0. \tag{3.11c}$$

Order ϵ^2:

$$\begin{aligned}\partial_0^2 u_2 + u_2'''' + k_1 u_2 = & -\partial_1^2 u_0 - 2\partial_0\partial_2 u_0 - 2\partial_0\partial_1 u_1 - \xi\,\partial_0 u_0 \\ & - 2k_2 u_0 u_1 - k_3 u_0^3 + \left(\frac{1}{2}\alpha q(s) e^{i\sigma_r t_2} e^{i\omega_r t_0} + \text{c.c.}\right),\end{aligned} \tag{3.12a}$$

$$u_2(0, t_k) = 0, \quad u_2'(0, t_k) = 0, \tag{3.12b}$$

$$u_2''(1, t_k) = 0, \quad -u_2'''(1, t_k) = \frac{1}{2}\alpha Q_B e^{i\sigma_r t_2} e^{i\omega_r t_0} + \text{c.c.}. \tag{3.12c}$$

Note that, in Eq. 3.12, the detuning entails a slow modulation, on the t_2 scale, of the harmonic excitation.

Solution to the Perturbation Equations

The generating Eqs. 3.10 constitute a linear eigenvalue problem governing the undamped free oscillations of the beam. They admit the general solution[4]:

$$u_0 = \sum_{k=1}^{\infty} a_k(t_1, t_2, \ldots)\phi_k(s)\cos(\omega_k t_0 + \varphi_k(t_1, t_2, \ldots)), \tag{3.13}$$

in which $\phi_k(s)$ is the kth natural mode, satisfying Eqs. 3.2, and a_k and φ_k are the kth amplitude and phase, respectively, which are modulated on the slower time scales $t_1, t_2, \ldots$ (since the equations only contain t_0-derivatives). Equation 3.13 collects the contribution of all the modes, and therefore the response is said to be *multimodal*. However, it can be expected that, *after a transient motion has been exhausted* and the beam reaches a steady-state regime,

[4] Here the symbols $\phi(s)$ and $\varphi(t)$ should not be confused, the first denoting a functions of the space, the latter of time.

the rth mode, which is involved in the primary resonance, supplies the most significant contribution to the response, if compared with the remaining non-resonant modes. Therefore, we will focus our attention on the steady-solution, and neglect in Eq. 3.13 all terms of the sum except the rth mode, by referring to it as a *monomodal solution*, namely[5]:

$$u_0 = a_r(t_1, t_2, \ldots)\phi_r(s)\cos(\omega_r t_0 + \varphi_r(t_1, t_2, \ldots)), \tag{3.14}$$

or:

$$u_0 = A_r(t_1, t_2, \ldots)\phi_r(s)e^{i\omega_r t_0} + \text{c.c.}, \tag{3.15}$$

where the *complex amplitude*:

$$A_r = \frac{1}{2}a_r(t_1, t_2, \ldots)e^{i\varphi_r(t_1, t_2, \ldots)} \tag{3.16}$$

has been introduced (which makes powering trigonometric expressions much more easier).

In order to attribute a physical meaning to the modulus of the amplitude, a_r, and to avoid arbitrary quantities ahead, we will require that the ω_r-harmonic component of the response, evaluated at $s = 1$, is equal to a_r. This entails that the mode must be normalized according to $\phi_r(1) = 1$, and moreover, the ω_r-harmonic of $u_j(1, t_0, t_1, \ldots)$ vanishes, when $j = 1, 2, \ldots$.[6]

Once Eq. 3.15 is substituted in the ϵ order perturbation Eqs. 3.11, these latter read:

$$\begin{aligned} \partial_0^2 u_1 + u_1'''' + k_1 u_1 = &-2i\omega_r \partial_1 A_r \phi_r e^{i\omega_r t_0} \\ &- k_2\phi_r^2\left(A_r^2 e^{2i\omega_r t_0} + A_r\bar{A}_r\right) + \text{c.c.}, \end{aligned} \tag{3.17a}$$

$$u_1(0, t_k) = 0, \quad u_1'(0, t_k) = 0, \tag{3.17b}$$

$$u_1''(1, t_k) = 0, \quad -u_1'''(1, t_k) = 0, \tag{3.17c}$$

where an overbar denotes the complex conjugate. These equations are of the same type of those studied in the Appendix A (Sect. A.4), i.e., linear equations with harmonic-varying know terms. Remarkably, we note that quadratic nonlinearities 'create' a new forcing frequency $2\omega_r$, plus a constant term. While this latter is surely non-resonant (since the space differential operator is non-singular, i.e., no buckling modes exist), in contrast, $2\omega_r$ is resonant if there exist a natural frequency $\omega_j = 2\omega_r$, as an eigenvalue of Eqs. 3.2. If this is the case, an *internal resonance* occurs among the rth and jth modes, entailing that this latter cannot be omitted in the generating solution. However, we will exclude this occurrence, so

---

5 If we took into account the complete Eq. 3.13, we would find that the amplitudes of the non-resonant modes exponentially decay in time.

6 Formally, the normalization condition reads $\int_0^{2\pi/\omega_r} e^{-i\omega_r t_0} u(1, t_0, t_1, \ldots)\,\mathrm{d}t_0 = a_r$. It entails $\int_0^{2\pi/\omega_r} e^{-i\omega_r t_0} u_0(1, t_0, t_1, \ldots)\,\mathrm{d}t_0 = a_r$ and $\int_0^{2\pi/\omega_r} e^{-i\omega_r t_0} u_j(1, t_0, t_1, \ldots)\,\mathrm{d}t_0 = 0$, when $j = 1, 2, \ldots$.

that, if $2\omega_r \neq \omega_j, \forall j$, the only resonant term in Eqs. 3.17a is the first one on the right hand side. For compatibility, this latter must be orthogonal to ϕ_r, from which it follows:

$$\partial_1 A_r = 0, \tag{3.18}$$

i.e., the unknown amplitude does not depend on the t_1-scale. A particular solution of Eqs. 3.17, consequently, reads:

$$u_1 = A_r^2 \chi_{12}(s) e^{2i\omega_r t_0} + A_r \bar{A}_r \chi_{10}(s) + \text{c.c.}, \tag{3.19}$$

where $\chi_{1m}(s)$, $m = 0, 2$, are solutions of the *non-singular* problems:

$$\chi_{1m}'''' + \left(k_1 - m^2\omega_r^2\right)\chi_{1m} = -k_2\phi_r^2(s), \tag{3.20a}$$

$$\chi_{1m}(0) = 0, \quad \chi_{1m}'(0) = 0, \tag{3.20b}$$

$$\chi_{1m}''(1) = 0, \quad -\chi_{1m}'''(1) = 0. \tag{3.20c}$$

The complementary solution of Eq. 3.17, which is $u_{1h} = B_r(t_1, t_2, \ldots)\phi_r(s)e^{i\omega_r t_0} + \text{c.c.}$, with B_r a new arbitrary amplitudes, must be ignored, according to the normalization adopted, which implies $B_r = 0$.[7]

By going on to the ϵ^2 order, and substituting previous results, the perturbation Eqs. 3.12 become:

$$\begin{aligned}\partial_0^2 u_2 + u_2'''' + k_1 u_2 &= q_{21}(s, t_2)e^{i\omega_r t_0} + q_{22}(s, t_2)e^{2i\omega_r t_0} \\ &\quad + q_{23}(s, t_2)e^{3i\omega_r t_0} + \text{c.c.},\end{aligned} \tag{3.21a}$$

$$u_2(0, t_k) = 0, \quad u_2'(0, t_k) = 0, \tag{3.21b}$$

$$u_2''(1, t_k) = 0, \quad -u_2'''(1, t_k) = \frac{1}{2}\alpha Q_B e^{i\sigma_r t_2} e^{i\omega_r t_0} + \text{c.c.}, \tag{3.21c}$$

where:

$$\begin{aligned}q_{21}(s, t_2) :&= -\partial_1^2 A_r \phi_r(s) - i\omega_r\left(\xi A_r + 2\partial_2 A_r\right)\phi_r(s) \\ &\quad - A_r^2 \bar{A}_r\left[2k_2(2\chi_{10}(s) + \chi_{12}(s))\phi_r(s) + 3k_3\phi_r^3(s)\right] + \frac{1}{2}\alpha q(s)e^{i\sigma_r t_2},\end{aligned} \tag{3.22a}$$

$$q_{22}(s, t_2) := -8i\omega_r A_r \partial_1 A_r \chi_{12}(s), \tag{3.22b}$$

$$q_{23}(s, t_2) := -A_r^3(k_3\phi_r^3(s) + 2k_2\chi_{12}(s)\phi_r(s)), \tag{3.22c}$$

[7] This explains the role of normalization: we do not want that higher order contributions of the response ‘repeat’ information already contained in the generating solution.

have the meaning of (order 2) space-dependent external excitation, pulsating with basic, double and triple frequency, respectively; the basic component, because of detuning, is also slowly modulated in time.

Equations 3.21 reveal the existence of a new possible internal resonance condition $3\omega_r = \omega_j$; by excluding its occurrence, compatibility only involves the ω_r-frequency forcing terms, and reads:

$$\int_0^1 \phi_r(s) q_{21}(s, t_2)\, \mathrm{d}s + \frac{1}{2}\phi_r(1)\alpha Q_B e^{i\sigma_r t_2} = 0, \tag{3.23}$$

from which the unknown $\partial_2 A_r$ is evaluated as:

$$\partial_2 A_r = -c_1 \xi A_r + i c_3 A_r^2 \bar{A}_r + i c_0 \alpha e^{i\sigma_r t_2}, \tag{3.24}$$

where the real constants c_k are defined below:

$$\begin{aligned} c_1 &:= \frac{1}{2}, \quad c_0 := -\frac{1}{4\omega_r \int_0^1 \phi_r^2(s)\mathrm{d}s}\left[\int_0^1 q(s)\phi_r(s)\, \mathrm{d}s + Q_B \phi_r(1)\right], \\ c_3 &:= \frac{3k_3 \int_0^1 \phi_r^4(s)\, \mathrm{d}s + 4k_2 \int_0^1 \phi_r^2(s)\chi_{10}(s)\, \mathrm{d}s + 2k_2 \int_0^1 \phi_r^2(s)\chi_{12}(s)\, \mathrm{d}s}{2\omega_r \int_0^1 \phi_r^2(s)\, \mathrm{d}s}. \end{aligned} \tag{3.25}$$

To come back to the original time, Eq. 3.8a must be taken into account, giving: $\dot{A}_r = \left(\partial_0 + \epsilon\partial_1 + \epsilon^2\partial_2 + \cdots\right) A_r$. Since A_r does not depend on t_0, and, for Eq. 3.18, $\partial_1 A_r = 0$, then, at this order, $\dot{A}_r = \epsilon^2 \partial_2 A_r$. Moreover, by multiplying Eq. 3.24 by ϵ^3, and using the inverse rescaling $\epsilon A_r \to A_r$, $\epsilon^3\alpha \to \alpha$, $\epsilon^2\sigma_r \to \sigma_r$, $\epsilon^2\xi \to \xi$, the perturbation parameter is reabsorbed, and Eq. 3.24 is rewritten in terms of the original variables as:

$$\dot{A}_r = -c_1 \xi A_r + i c_3 A_r^2 \bar{A}_r + i c_0 \alpha e^{i\sigma_r t}. \tag{3.26}$$

The Eq. 3.26 is the (complex) *bifurcation equation* of the problem at hand.[8]

Remark 3.1 It is observed that, since the $\chi_{1m}(s)$ functions are proportional to $-k_2$ (see Eq. 3.20), their contribution to c_3 is negative, irrespective of the sign of k_2. It follows that the *quadratic nonlinearities always induce a softening behavior* of the system (see the next discussion on Eq. 3.30 and the later Eq. 3.41). In contrast, the cubic nonlinearity k_3 can be either hardening ($k_3 > 0$) or softening ($k_3 < 0$).

Analysis of the bifurcation equation

Equation 3.26 can be split in two real equations, by using the polar representation of Eq. 3.16, and separating real and imaginary parts, thus obtaining:

[8] It is also known, in the nonlinear dynamic community, as the (complex) *Amplitude Modulation Equation* (AME). We, however, will prefer the former locution, used throughout the book.

$$\dot{a}_r = -c_1\xi a_r - 2c_0\alpha \sin\gamma_r, \tag{3.27a}$$

$$a_r\dot{\gamma}_r = a_r\sigma_r - \frac{1}{4}c_3 a_r^3 - 2c_0\alpha\cos\gamma_r, \tag{3.27b}$$

where:

$$\gamma_r := \sigma_r t - \varphi_r, \tag{3.28}$$

has been introduced to render the equations autonomous. Equations 3.27 will be referred to as the *real bifurcation equations.* They represent a *reduced two-dimensional dynamical system* which captures the essential dynamical behavior of the original infinite-dimensional system. In this respect, the MSM can be seen as a *reduction method*, as other existing in literature (see, e.g., [6]).

The meaning of the new variable γ_r clearly appears when we observe that, with the definitions of γ_r (Eq. 3.28) and of σ_r (Eq. 3.3), the total phase $\Phi_r(t) := \omega_r t + \varphi_r(t)$ of the generating solution in Eq. 3.14, becomes $\Phi_r(t) := \Omega t - \gamma_r(t)$. Therefore, γ_r is called the *phase-difference*, since it represents the phase-delay between the excitation and the (leading part) of the response.

Once Eqs. 3.27 have been integrated, may be numerically, the mono-modal response of the beam follows from the series expansion in Eq. 3.9 and the solutions at the various orders, (i.e., Eqs. 3.14, 3.19); it reads:

$$\begin{aligned} u &= a_r(t)\phi_r(s)\cos(\Omega t - \gamma_r(t)) \\ &+ \frac{1}{2}a_r^2(t)\{\chi_{12}(s)\cos[2(\Omega t - \gamma_r(t))] + \chi_{10}(s)\} + \cdots . \end{aligned} \tag{3.29}$$

Remark 3.2 Rescaling Eqs. 3.4 were aimed to shift excitation and damping at the ϵ^2 order, where *'significant' nonlinearities*, namely the cubic ones, appear for the first time in the relevant perturbation Eqs. 3.12. As a matter of fact, if excitation had been moved to the ϵ^0 order, the excitation would have appeared in the generating equations, leading to divergent solutions. If we had placed it at the ϵ^1 order, then the excitation would have appeared together with quadratic nonlinearities, which are unable to furnish competing ω_r-frequency terms, and again a divergent solution would have been obtained. Therefore, we rescaled the excitation to put it at the same level of the cubic nonlinearities which, indeed, provide ω_r-frequency terms. Damping, was consequently adjusted at the same order.

Remark 3.3 It is worth noticing that quadratic nonlinearities are responsible for the *change of the spatial shape* of oscillations, through the functions $\chi_{1m}(s)$. Therefore Eq. 3.29, even in the steady case, describes a motion in which the beam assumes different shapes in time, in contrast with the linear theory, in which a pattern (the natural mode) is fixed, and it just evolves with a scalar factor. This generalization of the linear mode is called in literature a *Nonlinear Normal Mode* (NNM, [7]). Of course, for these modes, the superposition principles, valid in the linear field, does not hold anymore. The NNM express the circumstance that an (active) linear mode (the rth, in our case) 'drives' all the remaining (passive) modes, which contribute to $\chi_{1m}(s)$. For these reasons, the $\chi's$ will be referred to as the *passive part of the response*.

Remark 3.4 The philosophies of the strained parameter method (here used in buckling analysis) and MSM (used in dynamics) are strongly different. In the former, the parameter(s) are expanded and compatibility equations are solved at each steps, to yield the unknown coefficients of the series. In the latter, parameters are not expanded, but only rescaled, and compatibility conditions *not* solved at each step, but rather reconstituted in a whole equation, which represents the reduced dynamical system. As an example, in more general cases than that analyzed here, if compatibility conditions of kind $\partial_1 A = f_1(A)$ and $\partial_2 A = f_2(A)$ are found at ϵ and ϵ^2 orders, respectively, they are reconstituted as $\dot{A} = f_1(A) + f_2(A)$. This procedure is called the *reconstitution method* [2, 4].

Periodic Motions and Stability

A very special, but important, case occurs when $a_r = a_{re} = \text{const}$, $\gamma_r = \gamma_{re} = \text{const}$, for which the previous equation describes a *periodic motion* of fundamental frequency Ω. Since these solutions are isolated, they represent *limit cycles*, which are traveled with periodic law. It is worth stressing that a periodic solution to the original equations of motion, is transformed by the MSM, in an *equilibrium point* for the reduced system. The strong simplification of the mathematical problem appears evident. However, it should be observed that this is *not* the general solution of the original problem, since it is monomodal.

To find the equilibrium points, we put $\dot{a}_r = \dot{\gamma}_r = 0$ in Eqs. 3.27 and look for solutions of two transcendental equations. The phase-difference, however, can be eliminated by squaring and summing the equations, to get a unique algebraic equation for the real amplitude:

$$4\alpha^2 c_0^2 = c_1^2 \xi^2 a_{re}^2 + \left(a_{re}\sigma_r - \frac{1}{4} c_3 a_{re}^3 \right)^2. \tag{3.30}$$

This is called the *frequency-response* law. Since it is a cubic equation for a_{re}^2, up to three real solutions are found for a given detuning. A typical plot for these roots is given in Fig. 3.1, both for damped (Fig. 3.1a, b) and undamped (Fig. 3.1c, d) beams. It is seen that the response is not infinite, as predicted by the linear theory, but the peak is finite and moves elsewhere with respect to the perfect resonance ($\sigma_r = 0$). Nonlinearities, indeed, bend the curves on the right (hardening effect, when $c_3 > 0$) or on the left (softening effect, when $c_3 < 0$), this entailing that the response is multivalued (so that *jumps* possibly occur while quasi-statically varying the external frequency). The stable or unstable character of the steady-solutions is also displayed in the figure, as determined by the following procedure.

To detect (orbital) stability of an equilibrium point (a_{re}, γ_{re}) of the reduced dynamical system in Eq. 3.27, we perturb it, by letting $a_r(t) = a_{re} + \delta a_r(t)$ and $\gamma_r(t) = \gamma_{re} + \delta\gamma_r(t)$,

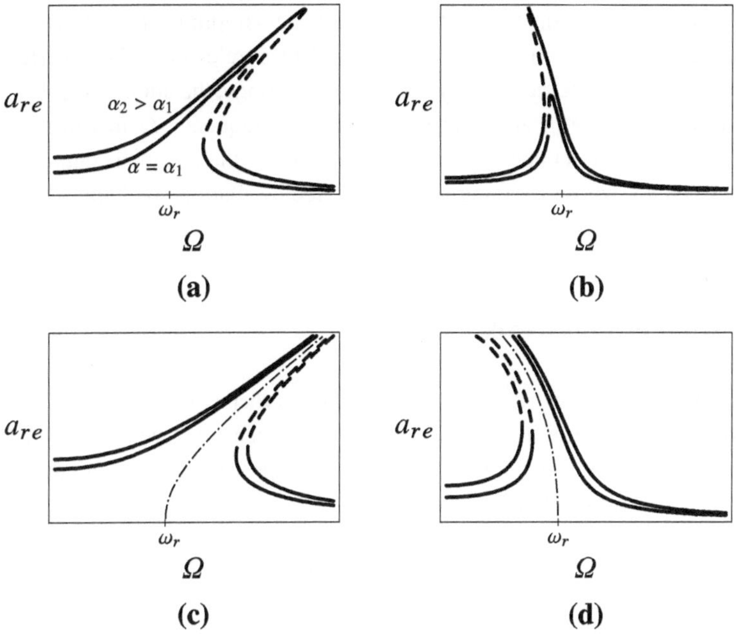

Fig. 3.1 Frequency-response law of damped **a, b** and undamped **c, d** systems in primary resonance: **a, c** hardening responses, **b, d** softening responses; continuous line: stable, dashed lines: unstable; dashed-dotted line: backbone curve

where $\delta a_r(t)$, $\delta\gamma_r(t)$ are small deviations from the point.[9] Substituting in the equation and linearizing in the perturbation, leads to the following *variational equation*[10]:

$$\begin{pmatrix} \delta\dot{a}_r \\ \delta\dot{\gamma}_r \end{pmatrix} = \begin{bmatrix} -c_1\xi & -2c_0\alpha\cos\gamma_{re} \\ 2c_0\frac{\alpha}{a_{re}^2}\cos\gamma_{re} - c_3\frac{a_{re}}{2} & 2c_0\frac{\alpha}{a_{re}}\sin\gamma_{re} \end{bmatrix} \begin{pmatrix} \delta a_r \\ \delta\gamma_r \end{pmatrix}. \tag{3.31}$$

This equation admits the solution $(\delta a_r(t) \;\; \delta\gamma_r(t)) = \mathbf{w}e^{\lambda t}$, where $(\lambda, \mathbf{w})$ is an eigenpair of $\mathbf{J}\mathbf{w} = \lambda\mathbf{w}$, and $\mathbf{J}$ is the 2×2 Jacobian matrix appearing in Eq. 3.31. It follows, that the equilibrium is stable if $\text{Re}(\lambda) < 0$ for all λ (since the perturbation decays in time), and unstable if $\text{Re}(\lambda) > 0$ for at least one λ (since the perturbation increases in time). It can be checked that, along the frequency-response curve, $\text{Re}(\lambda) = 0$ holds at the point whose tangent is vertical (and where a *saddle-node* bifurcation occurs, entailing a jump).

[9] Orbital stability is a more 'relaxed' property than stability, since it asks us only to check if the orbit is surrounded by sufficiently close orbits, without worrying about the way they are traveled, as instead Lyapunov stability requires. In other words, we say that an orbit is stable even if the point on it and the point on the perturbed orbit are, in some instant, very far away, provided the two orbits are close each other.

[10] An alternative procedure consists in applying the operator δ to Eq. 3.27, and to follow the rules of the formal *variational calculus*, from which the name of Eq. 3.31 derives.

Remark 3.5 It is interesting to note that the frequency-amplitude plot in Fig. 3.1c, d is formally identical to the bifurcation diagram in Fig. 2.3c, d, relevant to the imperfect pitchfork ruling buckling of symmetric static systems. In the analogy, the amplitude α of the external excitation plays the role of an imperfection, and the frequency Ω that of the bifurcation parameter. The 'imperfect curves' are thickened around the backbone curve, representing the perfect system, whose meaning as locus of free oscillations will appear clear soon, in Sect. 3.3.[11]

3.2.2 Symmetric Systems

When the beam is symmetric (i.e., $k_2 = 0$), the ϵ order perturbation Eq. 3.11 furnishes $u_1 = 0$, once the compatibility condition Eq. 3.18 has been enforced. We can take advantage from this circumstance and avoid trivial steps in the procedure, by a suitable rescaling of the variables. Thus, instead of Eqs. 3.4, we use:

$$u \to \epsilon^{1/2}\hat{u}, \quad \alpha \to \epsilon^{3/2}\hat{\alpha}, \quad \sigma_r \to \epsilon\hat{\sigma}_r, \quad \xi \to \epsilon\hat{\xi}, \tag{3.32}$$

(in which the new ϵ is the square of the old one). Moreover, we use the same power expansions Eqs. 3.8 and 3.9 for the variables and the differential operators, respectively, thus obtaining the following perturbation equations:

Order ϵ^0:

$$\partial_0^2 u_0 + u_0'''' + k_1 u_0 = 0, \tag{3.33a}$$

$$u_0(0, t_k) = 0 \quad u_0'(0, t_k) = 0, \tag{3.33b}$$

$$u_0''(1, t_k) = 0, \quad -u_0'''(1, t_k) = 0. \tag{3.33c}$$

Order ϵ^1:

$$\begin{aligned} \partial_0^2 u_1 + u_1'''' + k_1 u_1 &= -2\partial_0\partial_1 u_0 - \xi\partial_0 u_0 - k_3 u_0^3 \\ &\quad + \left(\frac{1}{2}\alpha q\,(s)\, e^{i\sigma_r t_1} e^{i\omega_r t_0} + \text{c.c.}\right), \end{aligned} \tag{3.34a}$$

$$u_1(0, t_k) = 0 \quad u_1'(0, t_k) = 0, \tag{3.34b}$$

$$u_1''(1, t_k) = 0, \quad -u_1'''(1, t_k) = \frac{1}{2}\alpha Q_B e^{i\sigma_r t_1} e^{i\omega_r t_0} + \text{c.c.}, \tag{3.34c}$$

in which only first order perturbations have been taken into account.

[11] The analogy would be more strict if we fixed Ω and varied ω as a free parameter of a family of beams. In this perspective, when ω equates Ω, a bifurcation from the trivial path $a = 0$ occurs, leading to the birth of natural oscillations on the backbone curve.

The monomodal generating solution is still given by Eq. 3.15; when it is used in Eq. 3.34, this latter reads:

$$\begin{aligned}\partial_0^2 u_1 + u_1'''' + k_1 u_1 = \Big[&-i\omega_r(\xi A_r + 2\partial_1 A_r)\phi_r(s)\\ &- 3k_3 A_r^2 \bar{A}_r \phi_r^3(s) + \frac{1}{2}\alpha q(s) e^{i\sigma_r t_1}\Big] e^{i\omega_r t_0}\\ &+ \text{c.c.} + \text{N.R.T.},\end{aligned} \tag{3.35a}$$

$$u_1(0, t_k) = 0 \quad u_1'(0, t_k) = 0, \tag{3.35b}$$

$$u_1''(1, t_k) = 0, \quad -u_1'''(1, t_k) = \frac{1}{2}\alpha Q_B e^{i\sigma_r t_1} e^{i\omega_r t_0} + \text{c.c.}, \tag{3.35c}$$

where N.R.T. is the acronym of Non-Resonant Terms. Compatibility then requires:

$$\begin{aligned}\int_0^1 \phi_r(s)\Big[&-i\omega_r(\xi A_r + 2\partial_1 A_r)\phi_r(s) - 3k_3 A_r^2 \bar{A}_r \phi_r^3(s)\\ &+ \frac{1}{2}\alpha q(s) e^{i\sigma_r t_1}\Big] \mathrm{d}s + \frac{1}{2}\phi_r(1)\alpha Q_B e^{i\sigma_r t_1} = 0,\end{aligned} \tag{3.36}$$

from which $\partial_1 A_r$ follows. By coming back to the true time and not rescaled variables, Eqs. 3.26–3.28 are recovered, in which c_0, c_1 are as in Eqs. 3.25, while c_3 is consistently updated as follows:

$$c_3 := \frac{3k_3}{2\omega_r} \frac{\int_0^1 \phi_r^4(s)\, \mathrm{d}s}{\int_0^1 \phi_r^2(s)\, \mathrm{d}s}. \tag{3.37}$$

The response of the system, at this order, remains monomodal, i.e., $u = a_r(t)\phi_r(s) \cos(\Omega t - \gamma_r(t))$. The frequency-response law, relevant to steady motions, is still given by Eq. 3.30, and plotted as in Fig. 3.1. The overall behavior is softening or hardening according to the sign of k_3 only.

Remark 3.6 In symmetric systems, the change of shape of the mode is an higher order effect. Therefore, the first order approximation of the frequency-response law only depends on the resonant (active) mode, not on the non-resonant (passive) modes.

3.3 Undamped Free Vibrations

It is well-known that, in the linear field, the *undamped*[12] natural frequencies are independent of the amplitude of oscillation. However, when finite amplitude are considered, the natural frequencies (and hence the periods) *do depend* on the amplitude. Here, we want to derive asymptotic formulas for these relationships.

We refer to an undamped non-symmetric beam, for which the equations of motion are a simplified version of Eqs. 3.1, i.e.:

$$\ddot{u} + u'''' + k_1 u + k_2 u^2 + k_3 u^3 + \ldots = 0, \tag{3.38a}$$

$$u(0, t) = 0, \quad u'(0, t) = 0, \tag{3.38b}$$

$$u''(1, t) = 0, \quad -u'''(1, t) = 0. \tag{3.38c}$$

By handling these equations by the MSM, exactly as we did for the forced and damped case, perturbation equations as Eqs. 3.10–3.12 are derived, in which $\alpha = 0$, $\xi = 0$. Hence, still excluding internal resonances, the compatibility condition Eq. 3.26 simplifies into:

$$\dot{A}_r = i c_3 A_r^2 \bar{A}_r. \tag{3.39}$$

This holds, of course, also for symmetric beams, when the expression in Eq. 3.37 is used for c_3, instead of Eq. 3.25. When Eq. 3.39 is put in the real form by the way of Eq. 3.16, it reads[13]:

$$\dot{a}_r = 0, \tag{3.40a}$$

$$a_r \dot{\varphi}_r = \frac{1}{4} c_3 a_r^3. \tag{3.40b}$$

The first of the previous equations only admits solutions $a_r = a_{re} = \text{const}$, denoting that the amplitude is always constant in time (i.e., no modulation occurs). This is due to the fact that there are neither sinks nor sources of energy in the system, so that the initial energy is stored.[14] Consequently, the second equation supplies a linear time-dependence for the phase, i.e., $\varphi_r = \Delta\omega_r t + \varphi_{r0}$, where:

$$\Delta\omega_r := \frac{1}{4} c_3 a_{re}^2 \tag{3.41}$$

[12] Damped free oscillations are not of interest, since they decay in time.

[13] Note that now there is no need to introduce a phase-difference, as, in contrast, occurs in the forced case.

[14] Therefore, all motions are periodic, and no limit cycles exist.

is called the *frequency correction*. Indeed, the total phase of the generating solution, $\Phi_r(t) := \omega_r t + \varphi_r(t)$, with the previous results, becomes $\Phi_r(t) := (\omega_r + \Delta\omega_r)t + \varphi_{r0} =: \varpi_r t + \varphi_{r0}$; here $\varpi_r := \omega_r + \Delta\omega_r$ is the rth *nonlinear frequency*, which depends on the squared amplitude a_{re}^2. The free response (to be compared with the forced response in Eq. 3.29), accordingly, is:

$$u = a_{re}\phi_r(s)\cos(\varpi_r t + \varphi_{r0}) + \frac{1}{2}a_{re}^2[\chi_{12}(s)\cos(2(\varpi_r t + \varphi_{r0})) + \chi_{10}(s)]. \quad (3.42)$$

It is worth noticing that, if we put $\xi = 0$ in the frequency-response law Eq. 3.30, and perform the limit for $\alpha \to 0$, this equations becomes identical to Eq. 3.41, when one identifies the frequency correction $\Delta\omega_r$ with the detuning σ_r. It is found, therefore, that the graph of the nonlinear frequency $\varpi_r = \varpi_r(a_{re})$ coincides with the frequency-response plot relevant to a vanishingly small harmonic excitation amplitude. This curve is called the *backbone*, and it is plotted in Fig. 3.1c, d. We can also said, by a reverse point of view, that the frequency-response curves of a forced undamped system are clustered around the backbone curve.

3.4 Sub- and Super-Harmonic Resonances

Two different forms of resonance, triggered by external excitation, are now examined. They are peculiar of the nonlinear behavior, and *do not exist in the linear field*. These resonances manifest when the excitation frequency Ω is a *multiple* or *sub-multiple* of a natural frequency ω_r. In the first case the resonance is called sub-harmonic (since ω_r is smaller than Ω), in the second super-harmonic (since ω_r is larger than Ω). It is important to stress that, in order for these resonances to give significant responses, since the linear response is non-resonant, *the amplitude α of the excitation must be large*. For this reason they are also called 'hard excitations', in contrast with the 'soft excitation' of the primary resonance. Here, the analysis will be confined to cubic (symmetric) systems, for which, at the lowest order, the resonances manifest when $\Omega \simeq 3\omega_r$ and $\Omega \simeq \frac{1}{3}\omega_r$, respectively.

3.4.1 Basic Mechanisms Leading to Nonlinear Resonances

The beam on elastic soil, already analyzed in the Sect. 3.2 (Eq. 3.1), is considered again, but with $k_2 = 0$; it is governed by[15]:

[15] A discrete version of the problem is dealt with in the Sect. B.4.2 of the Appendix B.

$$\ddot{u} + \xi \dot{u} + u'''' + k_1 u + k_3 u^3 + \cdots = \frac{1}{2} \alpha q(s) e^{i\Omega t} + \text{c.c.}, \tag{3.43a}$$

$$u(0, t) = 0, \quad u'(0, t) = 0, \tag{3.43b}$$

$$u''(1, t) = 0, \quad -u'''(1, t) = \frac{1}{2} \alpha Q_B e^{i\Omega t} + \text{c.c..} \tag{3.43c}$$

in which Ω, for the moment, is kept free to assume any value.

Perturbation Equations

The rescaling in Eq. 3.32 is still used, except for the amplitude of the excitation, namely, $u \to \epsilon^{1/2}\hat{u}$, $\xi \to \epsilon\hat{\xi}$, $\alpha \to \epsilon^{1/2}\hat{\alpha}$. Moreover, independent time scales $t_0 := t, t_1 := \epsilon t, \ldots$ are introduced, so that $\dot{u} = (\partial_0 + \epsilon \partial_1 + \cdots) u$ and $\ddot{u} = (\partial_0^2 + 2\epsilon \partial_0 \partial_1 + \cdots) u$. Finally, the displacement is expanded as $u = u_0 + \epsilon u_1 + \cdots$. The following perturbation equations are drawn:

Order ϵ^0:

$$\partial_0^2 u_0 + u_0'''' + k_1 u_0 = \frac{1}{2} \alpha q(s) e^{i\Omega t} + \text{c.c.}, \tag{3.44a}$$

$$u_0(0, t_k) = 0 \quad u_0'(0, t_k) = 0, \tag{3.44b}$$

$$u_0''(1, t_k) = 0, \quad -u_0'''(1, t_k) = \frac{1}{2} \alpha Q_B e^{i\Omega t} + \text{c.c..} \tag{3.44c}$$

Order ϵ^1:

$$\partial_0^2 u_1 + u_1'''' + k_1 u_1 = -2\partial_0 \partial_1 u_0 - \xi \partial_0 u_0 - k_3 u_0^3, \tag{3.45a}$$

$$u_1(0, t_k) = 0 \quad u_1'(0, t_k) = 0, \tag{3.45b}$$

$$u_1''(1, t_k) = 0, \quad -u_1'''(1, t_k) = 0. \tag{3.45c}$$

Solution

The solution to the generating perturbation Eq. 3.44 is made of two contributions: (i) the complementary solution, in which only the fundamental mode $\phi_r(s)$, involved in the resonance, is taken (the other ones being damped at the next orders), and (ii) a particular solution, i.e.:

$$u_0 = A_r(t_1, \ldots) \phi_r(s) e^{i\omega_r t_0} + \frac{1}{2} \alpha \chi_0(s) e^{i\Omega t_0} + \text{c.c.}, \tag{3.46}$$

in which χ_0 satisfies the following boundary value problem:

$$\chi_0'''' + (k_1 - \Omega^2)\chi_0 = q(s), \tag{3.47a}$$

$$\chi_0(0) = 0, \quad \chi_0'(0) = 0, \tag{3.47b}$$

$$\chi_0''(1) = 0, \quad -\chi_0'''(1) = Q_B. \tag{3.47c}$$

When Eq. 3.46 is substituted in the next order perturbation Eq. 3.45a, this becomes:

$$\begin{aligned}\partial_0^2 u_1 + u_1'''' + k_1 u_1 = {} & \underline{q_{11}(s) e^{i\omega_r t_0}} + q_{12}(s)\, e^{i\Omega t_0} + q_{13}(s)\, e^{3i\omega_r t_0} \\ & + \underline{q_{14}(s)\, e^{3i\Omega t_0}} + q_{15}(s)\, e^{i(\Omega+2\omega_r)t_0} \\ & + \underline{q_{16}(s)\, e^{i(\Omega-2\omega_r)t_0}} + q_{17}(s)\, e^{i(2\Omega+\omega_r)t_0} \\ & + q_{18}(s)\, e^{i(2\Omega-\omega_r)t_0} + \text{c.c.}.\end{aligned} \tag{3.48}$$

Among the excitation frequencies on the right hand member, since, by hypothesis, $\Omega \neq \omega_r$, there are only two frequencies which can be equal to $\pm\omega_r$ (in addition to ω_r itself), namely:

- $\Omega - 2\omega_r = \omega_r$, when $\Omega = 3\omega_r$ (sub-harmonic resonance);
- $3\Omega = \omega_r$, when $\Omega = \frac{1}{3}\omega_r$ (super-harmonic resonance).

Therefore, the only terms needing to be evaluated are those underlined in the previous equations, namely:

$$q_{11} := -2i\omega_r\phi_r\partial_1 A_r - i\omega_r\xi A_r\phi_r - 3k_3 A_r^2\bar{A}_r\phi_r^3 - \frac{3}{2}\alpha^2 k_3 A_r \chi_0^2\phi_r, \tag{3.49a}$$

$$q_{14} := -\frac{1}{8}\alpha^3 k_3 \chi_0^3, \tag{3.49b}$$

$$q_{16} := -\frac{3}{2}\alpha k_3 \bar{A}_r^2\phi_r^2\chi_0. \tag{3.49c}$$

The MSM, thus, reveals the mechanism leading to the two forms of nonlinear resonances. These latter are separately analyzed in the following sub-sections.

3.4.2 Sub-Harmonic Resonance

To analyze the sub-harmonic resonance, $\Omega = 3\omega_r + \epsilon\hat{\sigma}_r$ is introduced in the q_{16} term of Eq. 3.48 and the following compatibility condition enforced:

$$\int_0^1 \phi_r(s)\left[q_{11}(s) + q_{16}(s)\, e^{i\sigma_r t_1}\right] \mathrm{d}s = 0, \tag{3.50}$$

from which the bifurcation equation is drawn:

$$\dot{A}_r = -c_1 \xi A_r + i c_3 A_r^2 \bar{A}_r + i c_{21} \alpha^2 A_r + i c_{12} \alpha \bar{A}_r^2 e^{i \sigma_r t}, \tag{3.51}$$

in which:

$$c_1 := \frac{1}{2}, \tag{3.52a}$$

$$c_3 := \frac{3k_3 \int_0^1 \phi_r^4 (s) \, ds}{2\omega_r \int_0^1 \phi_r^2 (s) \, ds}, \tag{3.52b}$$

$$c_{21} := \frac{3k_3 \int_0^1 \phi_r^2 (s) \, \chi_0^2 (s) \, ds}{4\omega_r \int_0^1 \phi_r^2 (s) \, ds}, \tag{3.52c}$$

$$c_{12} := \frac{3k_3 \int_0^1 \phi_r^3 (s) \, \chi_0 (s) \, ds}{4\omega_r \int_0^1 \phi_r^2 (s) \, ds}. \tag{3.52d}$$

By letting $A_r = \frac{1}{2} a_r e^{i \varphi_r}$ in the bifurcation Eq. 3.51, and separating real and imaginary parts, one finds:

$$\dot{a}_r = -c_1 \xi a_r - \frac{\alpha}{2} c_{12} a_r^2 \sin \gamma_r, \tag{3.53a}$$

$$a_r \dot{\gamma}_r = \sigma_r a_r - \frac{3\alpha}{2} c_{12} a_r^2 \cos \gamma_r - \frac{3}{4} c_3 a_r^3 - 3\alpha^2 c_{21} a_r, \tag{3.53b}$$

where $\gamma_r := \sigma_r t - 3\varphi_r$. Consequently, the response of the beam, expressed at the leading order by Eq. 3.46, reads:

$$u = a_r(t) \phi_r(s) \cos \left[\frac{1}{3} \left(\Omega t - \gamma_r(t) \right) \right] + \alpha \chi_0 (s) \cos (\Omega t), \tag{3.54}$$

having used $\omega_r t + \varphi_r = \frac{1}{3} (\Omega t - \gamma_r(t))$. It highlights the presence of the $\frac{\Omega}{3}$ sub-harmonic of the excitation, from which the name of the resonance. It is stressed that $a_r(t)$ only affects a part of the solution.

The steady solutions are found by letting $\dot{a}_r = \dot{\gamma}_r = 0$; by eliminating γ_{re} between them, the frequency-response law is drawn:

$$\begin{aligned} &\frac{c_3^2}{4\alpha^2 c_{12}^2} a_{re}^4 - \left(1 + \frac{2c_3 \sigma_r}{3\alpha^2 c_{12}^2} - \frac{2c_3 c_{21}}{c_{12}^2} \right) a_{re}^2 \\ &+ \frac{4\sigma_r^2}{9\alpha^2 c_{12}^2} - \frac{8c_{21}\sigma_r}{3c_{12}^2} + \frac{4\alpha^2 c_{21}^2}{c_{12}^2} + \frac{4\xi^2 c_1^2}{\alpha^2 c_{12}^2} = 0. \end{aligned} \tag{3.55}$$

A typical plot of a_{re} *versus* Ω is represented in Fig. 3.2a. It is seen that nontrivial solutions for the a_{re}-dependent term do not cover all the possible values of external frequency. However,

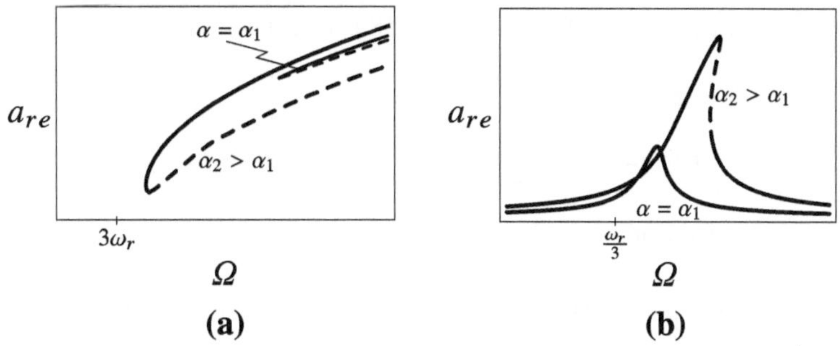

Fig. 3.2 Amplitude-frequency response law of systems under **a** sub-harmonic, **b** super-harmonic excitation; continuous line: stable, dashed line: unstable

frequencies outside the region of existence lead to the trivial solution, which is always stable, where the oscillations are ruled only by the higher-frequency particular solution (see [5] for further details).

To analyze stability of the steady solutions, the variational equation is built-up, which reads:

$$\begin{pmatrix} \delta \dot{a}_r \\ \delta \dot{\gamma}_r \end{pmatrix} = \begin{bmatrix} -c_1\xi - \alpha c_{12} a_{re} \sin\gamma_{re} & -\frac{1}{2}\alpha c_{12} a_{re}^2 \cos\gamma_{re} \\ -\frac{3}{2}c_3 a_{re} - \frac{3}{2}\alpha c_{12}\cos\gamma_{re} & \frac{3}{2}\alpha c_{12} a_{re}\sin\gamma_{re} \end{bmatrix} \begin{pmatrix} \delta a_r \\ \delta\gamma_r \end{pmatrix}. \tag{3.56}$$

The eigenvalues of the Jacobian matrix decides on stability. The relevant results are reported in Fig. 3.2a.

3.4.3 Super-Harmonic Resonance

To analyze the super-harmonic resonance, $3\Omega = \omega_r + \epsilon\hat{\sigma}_r$ is introduced in the q_{14} term of Eq. 3.48 and the following compatibility condition enforced:

$$\int_0^1 \phi_r(s)\left[q_{11}(s) + q_{14}(s)\, e^{i\sigma_r t_1}\right] \mathrm{d}s = 0, \tag{3.57}$$

from which the bifurcation equation is drawn:

$$\dot{A}_r = -c_1\xi A_r + i c_3 A_r^2 \bar{A}_r + i c_{21}\alpha^2 A_r + i c_{30}\alpha^3 e^{i\sigma_r t}, \tag{3.58}$$

where c_1, c_3 and c_{21} are still given by Eqs. 3.52, while:

$$c_{30} := \frac{k_3 \int_0^1 \phi_r(s)\, \chi_0^3(s)\, \mathrm{d}s}{16\omega_r \int_0^1 \phi_r^2(s)\, \mathrm{d}s}. \tag{3.59}$$

By letting $A_r = \frac{1}{2}a_r e^{i\varphi_r}$, the bifurcation Eq. 3.58 is recast as follows:

$$\dot{a}_r = -c_1\xi a_r - 2\alpha^3 c_{30}\sin\gamma_r, \tag{3.60a}$$

$$a_r\dot{\gamma}_r = \sigma_r a_r - \frac{1}{4}c_3 a_r^3 - \alpha^2 c_{21} a_r - 2\alpha^3 c_{30}\cos\gamma_r, \tag{3.60b}$$

where $\gamma_r := \sigma_r t - \varphi_r$. Consequently, at the leading order, the response of the beam, Eq. 3.46, reads:

$$u = a_r(t)\phi_r(s)\cos\left(3\Omega t - \gamma_r(t)\right) + \alpha\chi_0(s)\cos\left(\Omega t\right), \tag{3.61}$$

having used $\omega_r t + \varphi = 3\Omega t - \gamma_r(t)$. It reveals the existence of the 3Ω super-harmonic of the excitation, from which the name of the resonance. Once again, $a_r(t)$ governs only a part of the solution.

The steady solutions, for which $\dot{a}_r = \dot{\gamma}_r = 0$, satisfy the following frequency-response law:

$$\begin{aligned} &\frac{c_3^2}{64\alpha^6 c_{30}^2}a_{re}^6 + \left(\frac{c_3 c_{21}}{8\alpha^4 c_{30}^2} - \frac{c_3\sigma_r}{8\alpha^6 c_{30}^2}\right)a_{re}^4 \\ &+ \left(\frac{c_1^2\xi^2}{4\alpha^6 c_{30}^2} + \frac{c_{21}^2}{4\alpha^2 c_{30}^2} + \frac{\sigma_r^2}{4\alpha^6 c_{30}^2} - \frac{c_{21}\sigma_r}{2\alpha^4 c_{30}^2}\right)a_{re}^2 - 1 = 0. \end{aligned} \tag{3.62}$$

A qualitative plot of a_{re} *versus* Ω is represented in Fig. 3.2b. It appears that, differently from the sub-harmonic resonance, the super-harmonic steady motions exist for all the external frequencies. Multi-valued solutions for the a_{re}-dependent term may exist (e.g., for $\alpha = \alpha_2$ in the figure) and jump phenomena possibly occur as the external frequency is varied.

The stability of the steady solutions is ruled by the variational equation:

$$\begin{pmatrix}\delta\dot{a}_r\\ \delta\dot{\gamma}_r\end{pmatrix} = \begin{bmatrix} -c_1\xi & -2\alpha^3 c_{30}\cos\gamma_{re} \\ 2\alpha^3\frac{c_{30}}{a_{re}^2}\cos\gamma_{re} - \frac{1}{2}a_{re}c_3 & 2\alpha^3\frac{c_{30}}{a_{re}}\sin\gamma_{re}\end{bmatrix}\begin{pmatrix}\delta a_r\\ \delta\gamma_r\end{pmatrix}. \tag{3.63}$$

By analyzing the eigenvalues of the Jacobian matrix, the results reported in Fig. 3.2b are drawn.

3.5 Principal Parametric Excitation

A clamped-free beam is considered, resting on a Winkler soil with *symmetric* behavior,[16] axially loaded by a force keeping its direction, of nondimensional amplitude ν_d and pulsating with harmonic law of frequency Ω. The governing nondimensional equations of motion are drawn by Eqs. 1.7, with $\mu = \nu_s = 0$ and $\nu_d = \nu$, specialized as follows[17]:

---

[16] Here we will confine ourselves to the simpler case.

[17] A discrete version of the problem is dealt with in the Sect. B.5 of the Appendix B.

$$\ddot{u} + \xi \dot{u} + u'''' + 2\nu \cos(\Omega t)\ u'' + k_1 u + k_3 u^3 + \cdots = 0, \tag{3.64a}$$

$$u(0,t) = 0, \quad u'(0,t) = 0, \quad u''(1,t) = 0, \tag{3.64b}$$

$$-u'''(1,t) - 2\nu \cos(\Omega t)\, u'(1,t) = 0. \tag{3.64c}$$

3.5.1 Bifurcation Analysis

To tackle Eqs. 3.64, the MSM is applied. The rescaling $u \to \epsilon^{1/2}\hat{u}$, $\nu \to \epsilon\,\hat{\nu}$, $\xi \to \epsilon\,\hat{\xi}$ is introduced, which leads to (hat omitted):

$$\ddot{u} + u'''' + k_1 u + \epsilon\left[\xi \dot{u} + 2\nu \cos(\Omega t)\ u'' + k_3 u^3\right] = 0, \tag{3.65a}$$

$$u(0,t) = 0, \quad u'(0,t) = 0, \quad u''(1,t) = 0, \tag{3.65b}$$

$$-u'''(1,t) = 2\epsilon\, \nu \cos(\Omega t)\, u'(1,t). \tag{3.65c}$$

The displacement is expanded as $u = u_0 + \epsilon\, u_1 + \cdots$, and time-scales $t_0 := t$, $t_1 := \epsilon\, t$, $\cdots$ are introduced, for which $\partial_t = (\partial_0 + \epsilon\, \partial_1 + \cdots)$, $\partial_t^2 = (\partial_0 + \epsilon\, \partial_1 + \cdots)^2$. Hence, the following perturbation equations are derived:

Order ϵ^0:

$$\partial_0^2 u_0 + u_0'''' + k_1 u_0 = 0, \tag{3.66a}$$

$$u_0(0,t_k) = 0, \qquad u_0'(0,t_k) = 0, \tag{3.66b}$$

$$u_0''(1,t_k) = 0, \qquad -u_0'''(1,t_k) = 0. \tag{3.66c}$$

Order ϵ^1:

$$\begin{aligned} \partial_0^2 u_1 + u_1'''' + k_1 u_1 &= -2\partial_0\partial_1 u_0 - \xi\, \partial_0 u_0 - k_3 u_0^3 \\ &\quad - \nu\left(e^{i\Omega t_0} + \text{c.c.}\right) u_0'', \end{aligned} \tag{3.67a}$$

$$u_1(0,t_k) = 0, \ u_1'(0,t_k) = 0, \ u_1''(1,t_k) = 0, \tag{3.67b}$$

$$-u_1'''(1,t_k) = \nu\left(e^{i\Omega t_0} + \text{c.c.}\right) u_0'(1,t_k). \tag{3.67c}$$

From Eqs. 3.67 it follows that if u_0 contains the frequency ω_r, than $\Omega \simeq 2\omega_r$ produces a resonant term. Therefore, we let $\Omega = 2\omega_r + \epsilon\,\hat{\sigma}_r$, so that $e^{i\Omega t_0}$ reads as $e^{i\sigma_r t_1} e^{2i\omega_r t_0}$.

The solution to Eqs. 3.66, by accounting only for the rth natural mode ϕ_r involved in the resonance, reads[18]:

$$u_0 = A_r\,(t_1,\ldots)\,\phi_r\,(s)\,e^{i\omega_r t_0} + \text{c.c.}. \tag{3.68}$$

Consequently, Eqs. 3.67 become:

$$\partial_0^2 u_1 + u_1'''' + k_1 u_1 = -\Big(2i\omega_r \partial_1 A_r \phi_r + i\xi\omega_r A_r \phi_r + \nu \bar{A}_r e^{i\sigma_r t_1}\,\phi_r'' \tag{3.69a}$$

$$+ 3k_3 A_r^2 \bar{A}_r \phi_r^3\Big) e^{i\omega_r t_0} + \text{N.R.T.} + \text{c.c.},$$

$$u_1(0,t_k) = 0,\;\; u_1'(0,t_k) = 0,\;\; u_1''(1,t_k) = 0, \tag{3.69b}$$

$$-\,u_1'''(1,t_k) = \nu \bar{A}_r e^{i\sigma_r t_1} \phi_r'(1) e^{i\omega_r t_0} + \text{N.R.T.} + \text{c.c.}. \tag{3.69c}$$

Solvability requires that the resonant known term, in the domain and at the boundary, to be orthogonal to ϕ_r, i.e.:

$$-\int_0^1 \Big(2i\omega_r \partial_1 A_r \phi_r^2 + i\xi\omega_r A_r \phi_r^2 + \nu \bar{A}_r e^{i\sigma_r t_1}\,\phi_r \phi_r'' + 3k_3 A_r^2 \bar{A}_r \phi_r^4\Big)\,\mathrm{d}s$$
$$+ \nu \bar{A}_r e^{i\sigma_r t_1} \phi_r'(1)\phi_r(1) = 0, \tag{3.70}$$

from which $\partial_1 A_r$ is drawn. By multiplying the equation by $\epsilon^{3/2}$, and performing a backward rescaling via $\dot{A}_r = \epsilon\,\partial_1 A_r$, $\epsilon\,\xi \to \xi$, $\epsilon\,\nu \to \nu$, $\epsilon^{1/2} A_r \to A_r$, $t_1 \to \epsilon\, t$, $\epsilon\,\sigma_r \to \sigma_r$, the bifurcation equation is finally obtained:

$$\dot{A}_r = -c_1 \xi A_r + i c_3 A_r^2 \bar{A}_r + i c_0 \nu \bar{A}_r e^{i\sigma_r t}, \tag{3.71}$$

where the real constants are defined as follows[19]:

$$c_0 := \frac{1}{2\omega_r} \frac{\int_0^1 \phi_r \phi_r''\,\mathrm{d}s - \phi_r'(1)\phi_r(1)}{\int_0^1 \phi_r^2\,\mathrm{d}s},\quad c_1 := \frac{1}{2},\quad c_3 := \frac{3k_3}{2\omega_r} \frac{\int_0^1 \phi_r^4\,\mathrm{d}s}{\int_0^1 \phi_r^2\,\mathrm{d}s}. \tag{3.72}$$

Equation 3.71 governs the evolution of $A_r = A_r\,(t)$, which modulates the linear response in Eq. 3.68.

---

[18] The remaining modes give a smaller contribution to the response. If we considered one of them, e.g., of frequency $\omega_s \neq \omega_r$, the evolution of its amplitude a_s would be governed by the later bifurcation Eq. 3.73a, *deprived of the parametric excitation load*, i.e., $\dot{a}_s = -\frac{1}{2}\xi a_s$, stating that a_s decays in time.

[19] An integration by parts provides an alternative expression for c_0, i.e.:

$$c_0 := -\frac{1}{2\omega_r} \frac{\int_0^1 \phi_r'^2\,\mathrm{d}s}{\int_0^1 \phi_r^2\,\mathrm{d}s},$$

from which $c_0 < 0,\ \forall \phi_r$ is drawn.

Alternative Real Forms of the Bifurcation Equation

The complex bifurcation Eq. 3.71 is now rewritten in real form. By letting $A_r = \frac{1}{2}a_r(t)\,e^{i\varphi_r(t)}$, and separating the real and imaginary parts, two real bifurcation equations *in polar form* are derived, i.e.[20]:

$$\dot{a}_r = -c_1\xi a_r - c_0\,\nu\,a_r\,\sin\gamma_r, \tag{3.73a}$$

$$a_r\dot{\gamma}_r = \sigma_r a_r - \frac{1}{2}c_3 a_r^3 - 2c_0\,\nu\,a_r\,\cos\gamma_r, \tag{3.73b}$$

where the *phase difference*:

$$\gamma_r := \sigma_r t - 2\varphi_r \tag{3.74}$$

has been introduced to make the equations autonomous. Once Eqs. 3.73 have been integrated, possibly numerically, the first order solution $u \simeq u_0$ in Eq. 3.68 supplies[21]:

$$u = a_r(t)\,\phi_r(s)\cos\left(\frac{\Omega}{2}t - \frac{\gamma_r(t)}{2}\right). \tag{3.75}$$

An alternative form for the bifurcation equations, useful for future developments, is the *Cartesian form*.[22] First, however, Eq. 3.71 must be rendered autonomous. To this end, the change of variable $A_r = B_r e^{i\eta t}$, with η unknown, is introduced, and exponential functions dropped, leading to $\eta = \frac{\sigma_r}{2}$. Thus, the equation is transformed into:

$$\dot{B}_r + i\frac{\sigma_r}{2}B_r = -c_1\xi B_r + ic_3 B_r^2\bar{B}_r + ic_0\nu\bar{B}_r. \tag{3.76}$$

By letting $B_r = X_r + iY_r$ and separating the real and imaginary parts, the Cartesian form is found:

$$\dot{X}_r = -c_1\xi X_r + \left(c_0\nu + \frac{\sigma_r}{2}\right)Y_r - c_3\left(Y_r^3 + X_r^2 Y_r\right), \tag{3.77a}$$

$$\dot{Y}_r = -c_1\xi Y_r + \left(c_0\nu - \frac{\sigma_r}{2}\right)X_r + c_3\left(X_r^3 + X_r Y_r^2\right). \tag{3.77b}$$

---

[20] It should be noticed, that Eqs. 3.73 are similar but not equal to Eqs. 3.27, relevant to primary external resonance, since the unknown motion amplitude a replaces the know load amplitude α in the terms describing the source of the excitation.

[21] Indeed:

$$A_r\exp(i\omega_r t) = \frac{1}{2}a_r\exp\left[i\left(\omega_r t + \varphi_r\right)\right] = \frac{1}{2}a_r\exp\left[i\left(\omega_r + \frac{\sigma_r}{2}\right)t - i\frac{\gamma_r}{2}\right] = \frac{1}{2}a_r\exp\left(i\frac{\Omega}{2}t - i\frac{\gamma_r}{2}\right)$$

having used Eq. 3.74 and the resonance condition $\Omega = 2\omega_r + \sigma_r$ (ϵ reabsorbed).

[22] This is useful when the stability of the trivial solution $a_r = 0$ is checked, since the polar form is singular at this point.

After integration of Eqs. 3.77, the response reads[23]:

$$u = 2\left[X_r(t)\cos\left(\frac{\Omega}{2}t\right) - Y_r(t)\sin\left(\frac{\Omega}{2}t\right)\right]\phi_r(s). \tag{3.78}$$

3.5.2 Linear Stability Analysis of the Trivial Equilibrium

A linear stability analysis of the trivial equilibrium is preliminary carried out. Referring to the Cartesian form Eq. 3.77, and linearizing the motion around $(X_r, Y_r) = (0, 0)$, we get:

$$\begin{pmatrix} \dot{X}_r \\ \dot{Y}_r \end{pmatrix} = \begin{bmatrix} -c_1\xi & c_0\nu + \frac{\sigma_r}{2} \\ c_0\nu - \frac{\sigma_r}{2} & -c_1\xi \end{bmatrix}\begin{pmatrix} X_r \\ Y_r \end{pmatrix}. \tag{3.79}$$

By letting $(X_r, Y_r) = \left(\hat{X}_r, \hat{Y}_r\right)e^{\lambda t}$, an eigenvalue problem follows:

$$\begin{bmatrix} -c_1\xi - \lambda & c_0\nu + \frac{\sigma_r}{2} \\ c_0\nu - \frac{\sigma_r}{2} & -c_1\xi - \lambda \end{bmatrix}\begin{pmatrix} \hat{X}_r \\ \hat{Y}_r \end{pmatrix} = \begin{pmatrix} 0 \\ 0 \end{pmatrix}, \tag{3.80}$$

whose characteristic equation is:

$$\lambda^2 + 2c_1\xi\lambda + c_1^2\xi^2 - \left(c_0^2\nu^2 - \frac{\sigma_r^2}{4}\right) = 0, \tag{3.81}$$

supplying two eigenvalues $\lambda_{1,2}$. In order for the response in Eq. 3.77 not to diverge in time, $\mathrm{Re}\,(\lambda_k) \leq 0$ must hold for $k = 1, 2$.

The Undamped Case

The simplest undamped case ($\xi = 0$) is addressed first, for which $\lambda_{1,2} = \pm\sqrt{c_0^2\nu^2 - \frac{\sigma_r^2}{4}}$. It is $\mathrm{Re}\left(\lambda_{1,2}\right) = 0$ (neutral stable equilibrium) when $|\sigma_r| > 2\,|c_0|\,\nu$, and $\mathrm{Re}\left(\lambda_{1,2}\right) > 0$ (unstable equilibrium) when $|\sigma_r| < 2\,|c_0|\,\nu$. Therefore, there exist a region, on the (σ, ν) plane,[24] bounded (at this order of approximation) by two straight lines, $\sigma_r = \pm 2c_0\nu$, at which the trivial equilibrium of the beam loses stability. The internal region ($|\sigma_r|$ small) characterizes unstable systems, the external region ($|\sigma_r|$ large) stable systems. The region is plotted in Fig. 3.3a on the (Ω, ν) plane, by remembering that $\Omega = 2\omega_r + \sigma_r$.

[23] Indeed: $A_r \exp(i\omega_r t) = B_r \exp\left[i\left(\omega_r + \frac{\sigma_r}{2}\right)t\right] = B_r \exp\left(i\frac{\Omega}{2}t\right)$.

[24] This plane should be meant as the space of the bifurcation parameters (i.e., representative of the family of systems browsed) in which the bifurcation is unfolded.

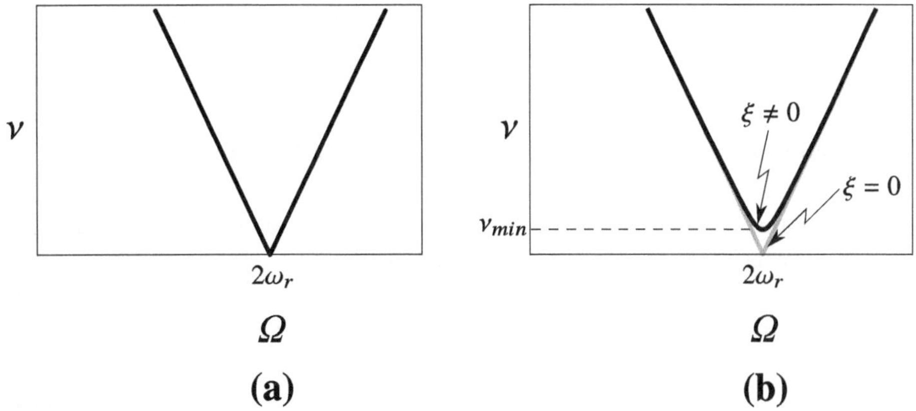

Fig. 3.3 Region of principal parametric resonance for **a** undamped, **b** damped systems

The Damped Case

When damping is present, i.e., $\xi \neq 0$, it is more convenient to determine the critical locus, in the (Ω, ν) plane, at which $\mathrm{Re}\left(\lambda_{1,2}\right) = 0$, where stability is incipiently lost. By letting $\lambda = i\beta$ in the characteristic Eq. 3.81, and separating real and imaginary parts, it follows:

$$-\beta^2 + c_1^2\xi^2 - \left(c_0^2\nu^2 - \frac{\sigma_r^2}{4}\right) = 0, \quad (3.82a)$$

$$2c_1\xi\beta = 0. \quad (3.82b)$$

From the second of these equations, $\beta = 0$ is drawn, i.e., both the eigenvalues λ vanish at the critical locus; from the first equation, the critical condition is found (for $\nu > 0$):

$$\nu = \frac{1}{|c_0|}\sqrt{c_1^2\xi^2 + \frac{\sigma_r^2}{4}}, \quad (3.83)$$

plotted in Fig. 3.3b. It is noticed that damping stabilizes the beam when $\nu < \nu_{min} := \frac{c_1\xi}{|c_0|}$ and slightly shrinks the unstable zone.

Remark 3.7 It is interesting to note that, without or with damping, it is $\lambda_{1,2} = 0$ at the critical locus of the parameter plane. Since the system of Eqs. 3.79 admits there the eigenvector $\hat{X}_r = \text{const}$, $\hat{Y}_r = \text{const}$, the associated linear response in Eq. 3.78 describes a harmonic motion of frequency $\frac{\Omega}{2}$, i.e., of *double period with respect to that of the parametric excitation*. This is consistent with the Floquet theory of the linear ordinary differential equations with periodic coefficients (see, e.g., [1, 5]), which states this property at the boundary of the principal resonance zone, at which a *flip bifurcation* takes place.[25]

3.5.3 Nonlinear Analysis of Bifurcated Periodic Motions

To perform a nonlinear analysis, the polar form of the governing bifurcation equations, Eqs. 3.73, is considered. First, the equilibrium points $a_r = a_{re} = \text{const}$, $\gamma_r = \gamma_{re} = \text{const}$ of the two dimensional dynamical system are sought for; they correspond (via Eq. 3.75) to periodic motions (limit cycles) for the beam, of frequency $\frac{\Omega}{2}$, i.e.:

$$u = a_{re}\,\phi_r(s)\cos\left(\frac{\Omega}{2}t - \frac{\gamma_{re}}{2}\right). \tag{3.84}$$

Steady amplitudes and phase differences satisfy the algebraic equations:

$$a_{re}\left(c_1\xi + c_0\nu\,\sin\gamma_{re}\right) = 0, \tag{3.85a}$$

$$a_{re}\left(\sigma_r - \frac{1}{2}c_3 a_{re}^2 - 2c_0\nu\,\cos\gamma_{re}\right) = 0. \tag{3.85b}$$

These admit the trivial solution $a_{re} = 0,\ \forall\gamma_{re}$,[26] and one or more nontrivial solutions. To evaluate them, the phase is eliminated by squaring and summing, to get:

$$\left(\frac{\sigma_r}{2} - \frac{1}{4}c_3 a_{re}^2\right)^2 + c_1^2\xi^2 = c_0^2\nu^2, \tag{3.86}$$

and the equation solved with respect the amplitude:

$$a_{re}^2 = \frac{4}{c_3}\left(\frac{\sigma_r}{2} \pm \sqrt{c_0^2\nu^2 - c_1^2\xi^2}\right). \tag{3.87}$$

This is the Cartesian equation of the surface $a_{re} = a_{re}(\sigma_r, \nu; \xi)$, with ξ a parameter. The surface intersects the critical locus belonging to the plane (σ_r, ν), from which it branches off.[27]

The undamped case is considered first, for which the amplitude locus of Eq. 3.87 reduces to $a_{re}^2 = \frac{4}{c_3}\left(\frac{\sigma_r}{2} \pm c_0\nu\right)$. It splits in two branches, $\mathcal{A}^{\pm}$, whose cross-sections $\nu = \text{const}$ are represented in Fig. 3.4a, b, in the hardening ($c_3 > 0$) and softening ($c_3 < 0$) cases,

---

[25] Therefore, the zeroing of the eigenvalue λ should *not* be erroneously meant as a divergence, since it refers to the complex amplitude $B_r = X_r + iY_r$, not to the response $u(s, t)$. As a matter of fact, in the framework of the MSM, the fast frequency ω_r, and successively the detuning $\frac{\sigma_r}{2}$, have been expunged from $u(s, t)$, to describe the slow motion $B_r(t)$.

[26] This circumstance entails the already cited singularity of the polar form, which leaves the phase undetermined.

[27] Indeed, when $a_{re} = 0$, Eq. 3.86 reduces to Eq. 3.83.

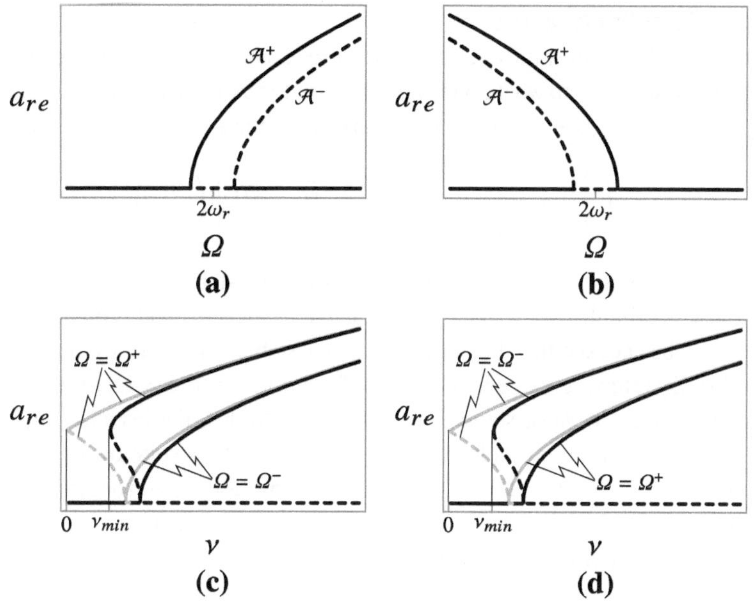

Fig. 3.4 Limit cycle amplitude, in principal parametric resonance: a_{re} *versus* Ω and fixed ν for undamped **a** hardening case, **b** softening case; a_{re} *versus* ν and fixed $\Omega = \Omega^{\pm} = 2\omega_r \pm \sigma_r$ for undamped (gray line) and damped (black line): **c** hardening case, **d** softening case. Continuous and dashed line: stable and unstable limit cycle, respectively

respectively (i.e., according to the sign of the cubic stiffness of the foundation soil of the beam). It is emerges that: (a) the limit cycle extends out of the instability region, on the right or on the left, respectively, and (b) the response is multivalued, this entailing the possible occurrence of jumps, when the frequency is quasi-statically varied. A different behavior is observed in the $\Omega = \text{const}$ plane, where the cases $\Omega < 2\omega_r$ and $\Omega > 2\omega_r$ must be discerned. By referring to the hardening case (Fig. 3.4c), it is seen that, when $\Omega > 2\omega_r$, the lower (unstable) branch $\mathcal{A}^-$ and the upper (stable) branch $\mathcal{A}^+$ are both crossed. However, they match each other at a cusp point when $\nu = 0$; also in this case a jump phenomena occurs, when ν is increased from zero. When, instead, $\Omega < 2\omega_r$, only the stable branch $\mathcal{A}^+$ is crossed. In the softening case (Fig. 3.4d), a similar behavior is observed, but with the two cases $\Omega < 2\omega_r$ and $\Omega > 2\omega_r$ exchanged.

When damping is different from zero, the amplitude locus of Fig. 3.4a, b has a similar form, but it exists only for sufficiently high excitation ν. Instead, the cusp point appearing in the Fig. 3.4c, d, transforms into a regular point with vertical tangent, occurring at $\nu = \nu_{min}$.

Limit Cycle Stability Analysis

To detect the stability of the limit cycles, the *variational equation* of the bifurcation Eqs. 3.73 must be built up, similarly to what done in Eq. 3.31 for the external resonance case. By letting $a_r = a_{re} + \delta a_r$, $\gamma_r = \gamma_{re} + \delta\gamma_r$ and then linearizing in the small increments, we obtain:

$$\begin{pmatrix} \delta\dot{a}_r \\ \delta\dot{\gamma}_r \end{pmatrix} = \begin{bmatrix} 0 & a_{re}\left(\frac{c_3}{4}a_{re}^2 - \frac{\sigma_r}{2}\right) \\ -c_3 a_{re} & -2c_1\xi \end{bmatrix} \begin{pmatrix} \delta a_r \\ \delta\gamma_r \end{pmatrix}. \tag{3.88}$$

The eigenvalues $\lambda_{1,2}$ of the 2×2 Jacobian matrix decide on stability, namely if $\mathrm{Re}\left(\lambda_{1,2}\right) < 0$ the limit cycle is orbitally stable (i.e., it is attractive of the surround orbits), otherwise it is unstable. It can be checked that the upper solution is stable, while the lower is unstable (see Exercise 3.3). By referring to the hardening (softening) case (Fig. 3.4a, b), if the frequency is increased (decreased) quasi-statically from the left (right), the response runs along the curve $\mathcal{A}^+$, even when the unstable zone has been crossed, in spite of the fact the trivial equilibrium is there stable. Moreover: (a) in the hardening case, a jump occurs when reducing the excitation frequency from the right, and (b) in the softening case, a jump occurs when increasing the excitation frequency from the left.

3.6 Exercises

The MSM, discussed in this chapter, is applied to solve some case studies relevant to the beam on elastic soil, namely: (i) the primary external resonance caused by harmonic transverse forces in 1:1 resonance with the fundamental frequency of the beam; (ii) the sub- and super-harmonic resonances, triggered by transverse forces in frequency ratios 3:1, 1:3 with the fundamental frequency; (iii) the principal parametric resonance, caused by axial loads pulsating with frequency ratio 2:1 with the fundamental frequency.

Exercise 3.1 (*Primary external resonance of a beam on elastic soil*) Let us consider a clamped-free beam on non-symmetric Winkler soil, loaded by loads $p(s,t) = \bar{q}\cos(\Omega t)$, $P_B \equiv 0$, with $\Omega = \omega_1 + \sigma_1$ close to the natural fundamental frequency of the beam, ω_1. The following numerical values of the parameters are taken: $k_1 = 3$, $k_2 = 6$, $k_3 = 15$, $\xi = 0.1$ and $\bar{q} = 1.0$. (a) Evaluate the fundamental frequency and mode of the beam. (b) Plot the frequency-response law.(c) Plot the steady time-evolution of the tip of the beam when $\alpha = 0.15$, $\sigma_1 = 0.21$.

(a) The first natural mode and frequency of the beam is evaluated by solving the eigenvalue problem Eq. 3.2. Since the solution depends on the parameter $\beta := \sqrt[4]{\omega_1^2 - k_1}$, its real or complex character must preliminary be discussed. To this end, we observe that if the beam were deprived of its bending stiffness, i.e., the term u'''' were neglected in Eq. 3.1, the natural frequency would become $\omega_1 = \sqrt{k_1}$ (being the nondimensional mass equal to 1). Therefore the presence of the bending stiffness increases the frequency, leading it to $\omega_1^2 > k_1$, which makes β a real number. Accordingly, the general solution of Eq. 3.2a is:

$$\phi(s) = a_1 \cos(\beta s) + a_2 \sin(\beta s) + a_3 \cosh(\beta s) + a_4 \sinh(\beta s). \tag{3.89}$$

Note that the elastic soil influences, by the way of β, both the natural frequency and the mode, although the form of the general integral in Eq. 3.89 is not affected by the soil.

By applying the boundary conditions in Eqs. 3.2, a homogeneous algebraic problem in the unknowns a_j is derived, of type $\mathbf{A}\mathbf{a} = \mathbf{0}$, where $\mathbf{a} = (a_1\ a_2\ a_3\ a_4)^T$. In order to get non-trivial solutions, the matrix $\mathbf{A}$ has to be made singular, so that the characteristic equation $\det \mathbf{A} := -2\beta^6(1 + \cos\beta \cosh\beta) = 0$ must be satisfied, or, by omitting the solution $\beta = 0$, $1 + \cos\beta\cosh\beta = 0$. The left hand member of this equation is plotted in Fig. 3.5a, from which it is found that the lower root is $\beta_1 = 1.8751$, corresponding to $\omega_1 = 3.919$ (confirming that $\omega_1^2 > k_1$). The associated mode $\phi_1(s)$ is show in Fig. 3.5b.

(b) The solutions $\chi_{1m}(s)$ of Eqs. 3.20 are successively found and plotted in Fig. 3.5b. Then, by using Eqs. 3.25, the numerical values assumed by the coefficients are drawn, as $c_1 = 0.5$, $c_0 = -0.10$, $c_3 = 2.824$. The frequency-response law provided by Eq. 3.30 is evaluated; it is plotted in Fig. 3.5c for different values of α, where the dashed line shows the unstable branch.

(c) When $\alpha = 0.15$ and $\sigma_1 = 0.21$ (i.e., when $\Omega = 4.13$), two stable equilibrium points for the Eqs. 3.27 are found from the plot in Fig. 3.5c, namely, the upper $a_{1e} = 0.567$, $\gamma_{1e} = 1.242$, and the lower $a_{1e} = 0.149$, $\gamma_{1e} = 2.890$. The response of the beam is provided by Eq. 3.29. In stationary condition, at the tip $s = 1$, it assumes the expression $u(1,t) = a_{1e}\cos(\Omega t - \gamma_{1e}) + \frac{a_{1e}^2}{2}\{0.093\cos[2(\Omega t - \gamma_{1e})] - 0.291\}$. By referring to the upper solution, the time evolution of the tip of the beam over a period is shown in Fig. 3.5d, where: (I) linear, and (II) quadratic solutions are distinguished in continuous and dashed lines, respectively, and are compared to the (N) numerical solution of the original Eqs. 3.1, represented in gray line. □

Exercise 3.2 (*Sub- and super harmonic external resonances of a beam on elastic soil*) Consider the same beam of Exercise 3.1, but: (i) on symmetric elastic soil (i.e., $k_2 = 0$), and (ii) with different values of the excitation frequency, namely: $\Omega = 3\omega_1 + \sigma_1$ and $3\Omega = \omega_1 + \sigma_1$. (a) Find the particular solution to the generating equation. (b, c) Plot the frequency-response law in the two resonance cases.

(a) The solution to the boundary value problem Eq. 3.47, with $q(s) = \bar{q}$ and $Q_B = 0$, by using the complementary solution in Eq. 3.89, reads:

$$\chi_0 = a_1 \cos(\beta s) + a_2 \sin(\beta s) + a_3 \cosh(\beta s) + a_4 \sinh(\beta s) + \frac{\bar{q}}{k_1 - \Omega^2}, \tag{3.90}$$

where $\beta := \sqrt[4]{\Omega^2 - k_1}$. By enforcing the boundary conditions, the constants are found to be:

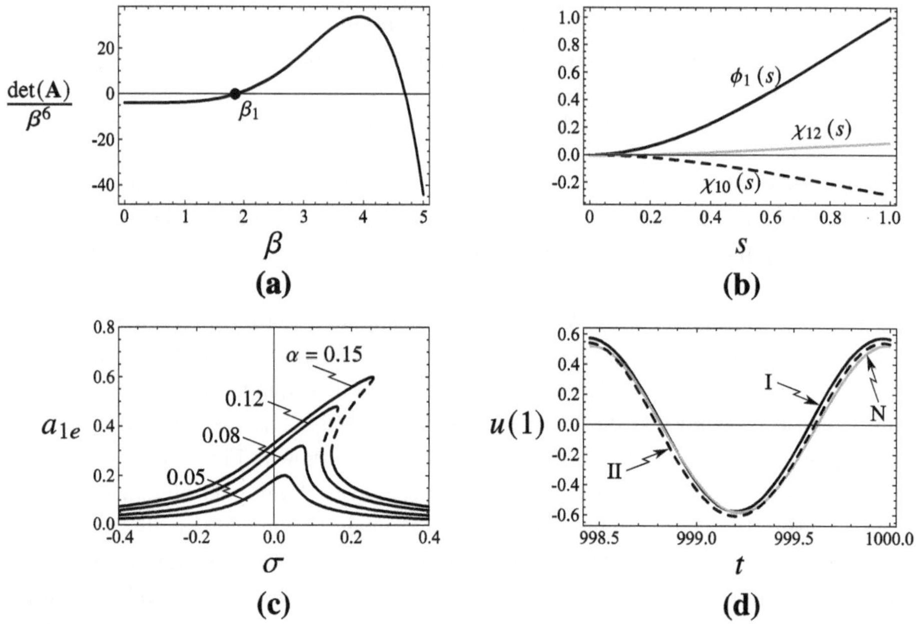

Fig. 3.5 Primary external resonance of the fundamental mode of the beam on elastic soil, when $k_1 = 3$, $k_2 = 6$, $k_3 = 15$, $\bar{q} = 1$: **a** determinant of the matrix $\boldsymbol{A}$ *versus* β; **b** fundamental natural mode (continuous line), passive contributions $\chi_{10}(s)$ (dashed line) and $\chi_{12}(s)$ (gray line); **c** frequency response curves (continuous line: stable; dashed line: unstable); **d** displacement of the tip of the beam, when $\sigma_1 = 0.21$ and $\alpha = 0.15$: (I) order ϵ^0 solution (continuous line); (II) order ϵ^1 solution (dashed line); (N) numeric solution (gray line)

$$a_1 = \frac{\bar{q}(1 + \cos\beta\cosh\beta - \sin\beta\sinh\beta)}{2\left(\Omega^2 - k_1\right)(1 + \cos\beta\cosh\beta)}, \qquad a_2 = \frac{\bar{q}(\cos\beta\sinh\beta + \sin\beta\cosh\beta)}{2\left(\Omega^2 - k_1\right)(1 + \cos\beta\cosh\beta)}, \tag{3.91a}$$

$$a_3 = \frac{\bar{q}(\sin\beta\sinh\beta + \cos\beta\cosh\beta + 1)}{2\left(\Omega^2 - k_1\right)(1 + \cos\beta\cosh\beta)}, \qquad a_4 = \frac{\bar{q}(\cos\beta\sinh\beta + \sin\beta\cosh\beta)}{2\left(k_1 - \Omega^2\right)(1 + \cos\beta\cosh\beta)}. \tag{3.91b}$$

(b) In the sub-harmonic case, by using the the previous results, the coefficients in Eqs. 3.52 are computed, resulting: $c_1 = 0.5$, $c_3 = 3.371$, $c_{21} = 0.00032$, $c_{12} = -0.023$. Then, the frequency-amplitude law in Eq. 3.55 is represented in Fig. 3.6a, for different values of α.

(c) In the super-harmonic case, The new coefficients in Eq. 3.59 are computed as $c_1 = 0.5$, $c_3 = 3.371$, $c_{21} = 0.0219$, $c_{12} = 0.00021$. The frequency-amplitude law in Eq. 3.62 is then plotted in Fig. 3.6b for different values of α. In Fig. 3.6, continuous and dashed lines shows the stable and unstable branches, respectively. □

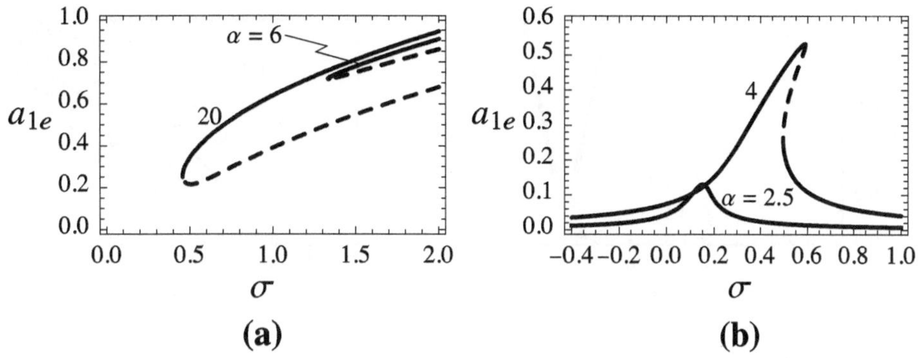

Fig. 3.6 Sub- and super-harmonic resonances of the fundamental mode of the beam on elastic soil, when $k_1 = 3, k_2 = 0, k_3 = 15, \bar{q} = 1$; frequency response curve in **a** sub-harmonic $\Omega = 3\omega_1$, and **b** super-harmonic $3\Omega = \omega_1$ resonances (continuous line: stable; dashed line: unstable)

Exercise 3.3 (*Principal parametric resonance of a beam on elastic soil*) A clamped-free beam resting on symmetric Winkler soil, loaded by a pulsating axial force in principal parametric resonance with the fundamental mode ($r = 1$), is considered, whose nondimensional equations of motion are given in Eqs. 3.64. Fix the following numerical values for the soil stiffness: $k_1 = 3, k_3 = 15$, and leaves the other parameters free. (a) Plot the linear stability domains for $\xi = (0, 0.1, 0.3)$. (b) Determine and plot the amplitude of the limit cycle versus the excitation frequency for the same values of damping and for $\nu = 0.27$. (c) Write the Jacobian matrix at selected equilibrium point, evaluate its eigenvalues, and show that the upper curve is stable and lower one is unstable.

(a) The fundamental eigenpair has already been found in the Exercise 3.1. The fundamental frequency is $\omega_1 = 3.919$ and the associated mode is plotted in Fig. 3.5. With this results, the coefficients in Eqs. 3.72 assume the values $c_0 = -0.593, c_1 = 0.5, c_3 = 3.371$. The stability region, in the undamped case, is bounded by the straight lines $\Omega = 2\omega_1 \pm 2c_0\nu$; in the damped case, by Eq. 3.83, all plotted in Fig. 3.7a.

(b) The amplitudes of the limit cycle are expressed by Eq. 3.87. For the selected values of damping, they are plotted in Fig. 3.7b, resulting in an upper branch $\mathcal{A}^+$ and in a lower branch $\mathcal{A}^-$. The upper branch extends on the two sides of the instability region, the lower branch only to the right.

(c) The stability of the cycle limit is governed by the eigenvalues $\lambda_{1,2}$ of the Jacobian matrix in Eq. 3.88. When the matrix is evaluated at the points $A^\pm$, $B^\pm$, ... on the branches $\mathcal{A}^\pm$, respectively, the eigenvalues reported in Table 3.1 are found. They establish that $\mathcal{A}^+$ is (orbitally) stable, and that $\mathcal{A}^-$ is unstable. □

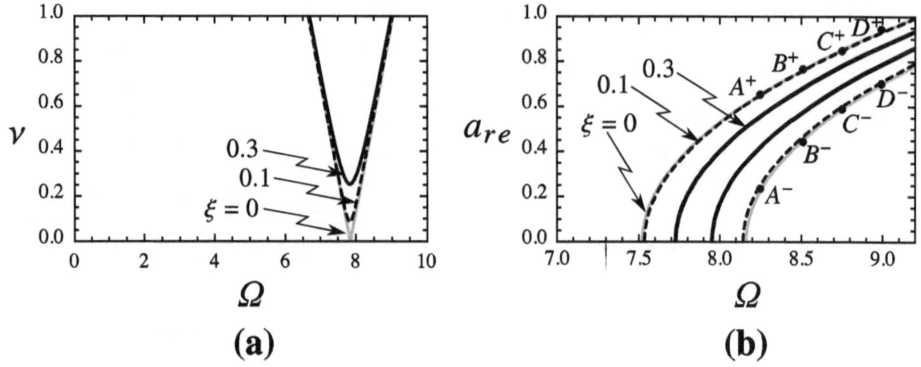

Fig. 3.7 Principal parametric excitation of the fundamental mode of the beam on Winkler soil, when $k_1 = 3, k_2 = 0, k_3 = 15$: **a** instability region when $\xi = (0, 0.1, 0.3)$; **b** amplitude of the limit cycle *versus* the excitation frequency, when $\xi = (0, 0.1, 0.3)$ and $\nu = 0.27$. Marked points $A^{\pm}$, $B^{\pm}$, ... denotes the limit cycles for which the stability analysis is carried out

Table 3.1 Eigenvalues of the Jacobian matrix in Eq. 3.88, computed at the points $A^{\pm}$, $B^{\pm}$, ... marked in Fig. 3.7b

	$\Omega = 8.25$				$\Omega = 8.5$			
	A^+		A^-		B^+		B^-	
ξ	a_{re}	$\lambda_{1,2}$	a_{re}	$\lambda_{1,2}$	a_{re}	$\lambda_{1,2}$	a_{re}	$\lambda_{1,2}$
0	0.66	$\pm 0.48\,i$	0.23	± 0.17	0.76	$\pm 0.56\,i$	0.45	± 0.33
0.1	0.65	$-0.05 \pm 0.46\,i$	0.25	$-0.24, 0.14$	0.76	$-0.05 \pm 0.54\,i$	0.46	$-0.38, 0.28$
0.3	0.56	$-0.15 \pm 0.19\,i$	0.42	$-0.39, 0.09$	0.67	$-0.15 \pm 0.25\,i$	0.57	$-0.44, 0.14$
	$\Omega = 8.75$				$\Omega = 9$			
	C^+		C^-		D^+		D^-	
ξ	a_{re}	$\lambda_{1,2}$	a_{re}	$\lambda_{1,2}$	a_{re}	$\lambda_{1,2}$	a_{re}	$\lambda_{1,2}$
0	0.85	$\pm 0.63\,i$	0.59	± 0.43	0.94	$\pm 0.69\,i$	0.71	± 0.52
0.1	0.85	$-0.05 \pm 0.61\,i$	0.60	$-0.48, 0.38$	0.93	$-0.05 \pm 0.67\,i$	0.71	$-0.56, 0.46$
0.3	0.78	$-0.15 \pm 0.30\,i$	0.69	$-0.48, 0.18$	0.86	$-0.15 \pm 0.35\,i$	0.79	$-0.52, 0.22$

References

1. Luongo, A., Ferretti, M., Di Nino, S.: Stability and Bifurcation of Structures: Statical and Dynamical Systems. Springer, Cham (2023)
2. Luongo, A., Paolone, A.: On the reconstitution problem in the multiple time-scale method. Nonlinear Dynamics **19**(2), 135–158 (1999)
3. Nayfeh, A.H.: Perturbation methods. John Wiley & Sons, New York (1973)
4. Nayfeh, A.H.: 'Topical course on nonlinear dynamics' in Perturbation Methods in Nonlinear Dynamics. Societa Italiana di Fisica, Santa Margherita di Pula, Sardinia (1985)

5. Nayfeh, A.H., Mook, D.T.: Nonlinear oscillations. Wiley, New York (1995)
6. Steindl, A., Troger, H.: Methods for dimension reduction and their application in nonlinear dynamics. International Journal of Solids and Structures **38**(10–13), 2131–2147 (2001)
7. Vakakis, A.F.: Normal Modes and Localization in Nonlinear Systems. Springer, Dordrecht (2011)

4 Autonomous Nonconservative Dynamical Systems

4.1 Introduction

Elastic structures loaded by forces which do not admit a potential are nonconservative systems. If these forces do not depend explicitly on time, then the system is autonomous. Examples of non-conservative forces are represented by the *follower forces*, which keep their modulus constant in time, but change their direction depending on the configuration of the elastic body. Here, the sample system previously introduced, made of a beam on Winkler soil, is considered, loaded by a compressive force applied at one end, whose instantaneous direction coincides with the tangent to the beam axis at that point. The problem is successively generalized by accounting for the simultaneous presence of a dead compressive load.

At critical values of these forces, taken as bifurcation parameters, dynamic (Hopf) or static (divergence) bifurcations manifest themselves, occurring at bifurcation loci represented by curves on the parameter plane. A family of limit cycles originate at a Hopf bifurcation point, whose amplitude and frequency depend on the magnitude of the load; a family of non-trivial equilibria bifurcate from a divergence point. Here, the analysis is carried out by the *Multiple Scale Method*, which reduces the infinite dimensional system to a two-dimensional one, in case of dynamic bifurcation, and to a one-dimensional one, in case of static bifurcation. The method allows investigating the existence of steady solutions in the postcritical range, together with their stability.

4.2 Dynamic Bifurcations

We want to analyze the behavior of the beam on elastic soil when it undergoes a *dynamic bifurcation*. This phenomenon only manifests in nonconservative systems, when a *positional force* exists, able to render complex, with *positive* real part, the eigenvalues of the associated

A. Luongo et al., *Perturbation Methods and Nonlinear Phenomena*,
Synthesis Lectures on Engineering, Science, and Technology,
https://doi.org/10.1007/978-3-031-49397-3_4

linear system.[1] At low values of the force, the beam still experiences harmonic motions, as a conservative system; however, when a threshold value of the force is reached, it manifests oscillations of amplitude exponentially increasing in time. If one confines himself to the linear theory, there is no way to arrest the amplitude divergence; however, if nonlinearities are taken into account, one could find that they limit the oscillation amplitude. We want to investigate how nonlinearity work, and which is the maximum amplitude reached. Dynamic bifurcations are known as *Hopf bifurcations* in Dynamical System Theory, and *flutter* (when damping is absent) or *galloping* (when damping is present) in the engineering field.

Remark 4.1 To explain the phenomenon in a simple way, let us consider the free motion $\mathbf{x}(t)$ of a linear discrete system. It is ruled by the ordinary differential equations $\ddot{\mathbf{x}} + \mathbf{K}\mathbf{x} = \mathbf{0}$, where $\mathbf{M} = \mathbf{I}$ has been taken[2] and $\mathbf{K} \neq \mathbf{K}^T$. In order for the motion to be harmonic of frequency ω, i.e., $\mathbf{x}(t) = \mathbf{u}e^{i\omega t}$, the algebraic eigenvalue problem $(\mathbf{K} - \Lambda\mathbf{I})\,\mathbf{u} = \mathbf{0}$ must admits *real and positive roots* $\Lambda := \omega^2$, entailing *real frequencies* $\omega_{1,2} = \pm\sqrt{\Lambda}$. This is true when $\mathbf{K}$ is symmetric and definite positive (i.e., the system is conservative and stable), but it is not assured when $\mathbf{K}$ is generic. Therefore, so far the nonconservative action is weak, $\Lambda > 0$ holds. However, when the nonconservative force is sufficiently large, a couple of eigenvalues of $\mathbf{K}$ becomes complex conjugate, i.e., $\Lambda_{1,2} = \rho e^{\pm i\theta}$, entailing the existence a quadruplet of *complex frequencies* $\omega_{1,2,3,4} = \pm\sqrt{\rho}e^{\pm i\frac{\theta}{2}}$. Therefore, one of the components of motion exponentially diverges to infinity with a complex characteristic exponent, $e^{(a+ib)t}$ $(a > 0, b > 0)$.

Remark 4.2 Referring to the previous example, it is of interest evaluating the power $\mathcal{P}$, expended, in a harmonic motion $\mathbf{x}(t) = \mathbf{u}\cos(\omega t)$, by the forces $\mathbf{f} = -\mathbf{K}_a\mathbf{x}$, where $\mathbf{K}_a = -\mathbf{K}_a^T$ is the antisymmetric part of $\mathbf{K}$, responsible for nonconservativeness. It turns out that, due to the antisimmetry of $\mathbf{K}_a$, it is $\mathcal{P} := \dot{\mathbf{x}}^T\mathbf{f} = \frac{1}{2}\omega\,\mathbf{u}^T\mathbf{K}_a\mathbf{u}\sin(2\omega t) = 0\;\forall t$, i.e., *the nonconservative forces are orthogonal to the velocity*. This results explains the reason for which a weakly nonconservative system can experience harmonic motions, behaving as it were conservative. However, when nonconservativeness are large, the motion becomes exponential, and $\mathbf{u}$ complex, i.e., $\mathbf{x}(t) = \frac{1}{2}\mathbf{u}e^{(a+ib)t} + \text{c.c.}$. Therefore $\mathcal{P} := -\frac{1}{4}\left((a+ib)\,\mathbf{u}e^{(a+ib)t} + \text{c.c.}\right)^T \mathbf{K}_a\left(\mathbf{u}e^{(a+ib)t} + \text{c.c.}\right)$, i.e., $\mathcal{P} = i\frac{b}{2}e^{2at}\bar{\mathbf{u}}^T\mathbf{K}_a\mathbf{u} \gtrless 0$, denoting that energy is pumped into or expunged from the system.[3]

---

[1] A similar case occurs when the force is velocity-dependent, as in aeroelasticity [3]. This aspects, however, will be not examined in this book.

[2] It is always possible to make the mass matrix unitary, by performing a suitable change of coordinates.

[3] The term $\bar{\mathbf{u}}^T\mathbf{K}_a\mathbf{u}$ is a purely imaginary number. Indeed, denoting with $\mathbf{u}_R$ and $\mathbf{u}_I$ the real and the imaginary part of $\mathbf{u}$, it is:

$$\begin{aligned}\bar{\mathbf{u}}^T\mathbf{K}_a\mathbf{u} &= (\mathbf{u}_R - i\mathbf{u}_I)^T\,\mathbf{K}_a\,(\mathbf{u}_R + i\mathbf{u}_I) = \cancel{\mathbf{u}_R^T\mathbf{K}_a\mathbf{u}_R} - i^2\cancel{\mathbf{u}_I^T\mathbf{K}_a\mathbf{u}_I} - i\mathbf{u}_I^T\,\mathbf{K}_a\mathbf{u}_R + i\mathbf{u}_R^T\mathbf{K}_a\mathbf{u}_I \\ &= i\left(\mathbf{u}_R^T\mathbf{K}_a\mathbf{u}_I - \mathbf{u}_I^T\mathbf{K}_a\mathbf{u}_R\right) = i\left(\mathbf{u}_R^T\mathbf{K}_a\mathbf{u}_I - \mathbf{u}_R^T\mathbf{K}_a^T\mathbf{u}_I\right) = 2\,i\,\mathbf{u}_R^T\mathbf{K}_a\mathbf{u}_I\end{aligned}$$

4.2.1 Non-symmetric Systems

We consider the same beam studied in Sect. 2.3, but prestressed by a follower (tangential) force, instead of a gravitational force. Since the problem is no more static but dynamic, we have to account for time dependence of the response $u(s,t)$. By letting $\nu_s = \nu_d = \alpha = 0$, in Eqs. 1.7, the equations of motion are derived[4]:

$$\ddot{u} + \xi\dot{u} + u'''' + 2\mu u'' + k_1 u + k_2 u^2 + k_3 u^3 + \cdots = 0, \tag{4.1a}$$

$$u(0,t) = 0, \quad u'(0,t) = 0, \tag{4.1b}$$

$$u''(1,t) = 0, \quad -u'''(1,t) = 0. \tag{4.1c}$$

Rescaling

First, we have to perform a rescaling. We order the displacement as $u \to \epsilon\hat{u}$ but, although the damping coefficient is often small, we do *not* rescale ξ, by assuming it is not evanescent, to avoid degenerateness of the eigenvalues at the bifurcation.[5] Moreover, we split the parameter μ in two parts: (a) a *critical* value μ_0, selecting the (unknown) bifurcating system, and, (b) an *incremental* part $\tilde{\mu}$, able to explore the neighborhood of bifurcation[6]:

$$\mu = \mu_0 + \tilde{\mu}. \tag{4.2}$$

Finally, we order the incremental parameter as $\tilde{\mu} \to \epsilon^2\tilde{\mu}$, since we know that the significant nonlinearities are the cubic ones. Thus, the equations of motion are rescaled as follows, after division by ϵ:

---

with $\mathbf{u}_R^T \mathbf{K}_a \mathbf{u}_I \gtrless 0$.

[4] A discrete version of the problem is dealt with in the Sect. B.6 of the Appendix B.

[5] When damping is zero (*circulatory* system), the mechanism of bifurcation requires a collision of two pairs of eigenvalues running on the imaginary axis, with subsequent separation of the eigenvalues in the two half-planes, stable and unstable; therefore the eigenvalue, at bifurcation, has multiplicity 2, and perturbation analysis calls for special treatments. A small damping, however, destroys this mechanism, since the two pairs approach each other in the left half-plane of the complex plane and then rapidly veer in opposite direction; therefore only one pair crosses the imaginary axis, making the critical eigenvalue of multiplicity 1. Said in other words, damping renders the bifurcation *generic*, while bifurcations of circulatory systems are *non-generic*.

[6] Here $\tilde{\mu}$ plays the same role of the detuning in the resonant excitation problem. Note that this splitting is somewhat different from the series expansion of the parameter, as the one we performed, e.g., in Eq. 2.18b, since $\tilde{\mu}$ has to be meant now as a known quantity (bifurcation parameter).

$$\ddot{u} + \xi \dot{u} + u'''' + 2\mu_0 u'' + k_1 u + \epsilon k_2 u^2 + \epsilon^2 \left(2\tilde{\mu} u'' + k_3 u^3\right) + \cdots = 0, \tag{4.3a}$$

$$u(0,t) = 0, \quad u'(0,t) = 0, \tag{4.3b}$$

$$u''(1,t) = 0, \quad -u'''(1,t) = 0. \tag{4.3c}$$

Multiple Scale Method

We tackle the problem by using the MSM. By exploiting the results obtained in the previous chapter, and in order to reduce the calculations, we skip the odd time scales because, again, the unknown amplitude would be independent of them. Accordingly, we expand the variable as:

$$u(s, t_0, t_2, \ldots) = u_0(s, t_0, t_2, \ldots) + \epsilon u_1(s, t_0, t_2, \ldots) + \epsilon^2 u_2(s, t_0, t_2, \ldots) + \cdots, \tag{4.4}$$

where $t_k := \epsilon^k t$, and the time differential operators as:

$$\partial_t = \partial_0 + \epsilon^2 \partial_2 + \cdots, \qquad \partial_t^2 = \partial_0^2 + 2\epsilon^2 \partial_0 \partial_2 + \cdots, \tag{4.5}$$

where $\partial_k := \partial_{t_k}$, $k = 0, 2, \ldots$. By substituting these expressions in the rescaled Eqs. 4.3, and separately equating to zero the terms with the same power of ϵ, it follows:

Order ϵ^0:

$$\partial_0^2 u_0 + \xi \partial_0 u_0 + u_0'''' + 2\mu_0 u_0'' + k_1 u_0 = 0, \tag{4.6a}$$

$$u_0(0, t_k) = 0, \quad u_0'(0, t_k) = 0, \tag{4.6b}$$

$$u_0''(1, t_k) = 0, \quad -u_0'''(1, t_k) = 0. \tag{4.6c}$$

Order ϵ^1:

$$\partial_0^2 u_1 + \xi \partial_0 u_1 + u_1'''' + 2\mu_0 u_1'' + k_1 u_1 = -k_2 u_0^2, \tag{4.7a}$$

$$u_1(0, t_k) = 0, \quad u_1'(0, t_k) = 0, \tag{4.7b}$$

$$u_1''(1, t_k) = 0, \quad -u_1'''(1, t_k) = 0. \tag{4.7c}$$

Order ϵ^2:

$$\begin{aligned}\partial_0^2 u_2 + \xi \partial_0 u_2 + u_2'''' + 2\mu_0 u_2'' + k_1 u_2 &= -2\partial_0 \partial_2 u_0 - \xi \partial_2 u_0 \\ &\quad - 2\tilde{\mu} u_0'' - 2k_2 u_0 u_1 - k_3 u_0^3,\end{aligned} \tag{4.8a}$$

$$u_2(0, t_k) = 0, \quad u_2'(0, t_k) = 0, \tag{4.8b}$$

$$u_2''(1, t_k) = 0, \quad -u_2'''(1, t_k) = 0. \tag{4.8c}$$

Right and Left Eigenvectors

The generating Eq. 4.6 is a not-self adjoint eigenvalue problem, both in space and time. By letting $u_0 = \phi_k(s)e^{\lambda_k t_0}$, a space-boundary value problem for the eigenpair (λ_k, ϕ_k), $k = 1, 2, \ldots$, follows:

$$\phi_k'''' + 2\mu_0\phi_k'' + \left(k_1 + \xi\lambda_k + \lambda_k^2\right)\phi_k = 0, \tag{4.9a}$$

$$\phi_k(0) = 0, \quad \phi_k'(0) = 0, \tag{4.9b}$$

$$\phi_k''(1) = 0, \quad -\phi_k'''(1) = 0. \tag{4.9c}$$

By keeping fixed the damping ξ and the linear soil stiffness k_1 (auxiliary parameters), but varying the prestress μ_0 (bifurcation parameter), the eigenvalues also vary, i.e., $\lambda_k = \lambda_k(\mu_0)$. For small values of the prestress, the eigenvalues are complex with real negative parts, i.e., the free motion is damped, and the beam, when initially disturbed, experiences damped oscillations, coming back to the rectilinear equilibrium position. However, when the prestress is increased, a pair of eigenvalues possibly cross the imaginary axis, from the left to the right, of the complex plane, causing a Hopf bifurcation. We assume that a critical value μ_c of μ_0 exists, such that a pair of eigenvalues is purely imaginary, $\lambda_c := \lambda(\mu_c) = \pm i\omega_c$ (and, moreover $\left(\partial_{\mu_0}\lambda_c\right)_{\mu_0=\mu_c} > 0$, said *transversality condition*). Denoting with ϕ_c the associated eigenvector (critical mode), which is in general a *complex* quantity, Eqs. 4.9 specialize into:

$$\phi_c'''' + 2\mu_c\phi_c'' + \left(k_1 + i\xi\omega_c - \omega_c^2\right)\phi_c = 0, \tag{4.10a}$$

$$\phi_c(0) = 0, \quad \phi_k'(0) = 0, \tag{4.10b}$$

$$\phi_k''(1) = 0, \quad -\phi_k'''(1) = 0. \tag{4.10c}$$

These equations are a particular case of Eqs. A.20, so that the adjoint homogeneous problem follows from Eqs. A.22:

$$\psi_c'''' + 2\mu_c\psi_c'' + \left(k_1 - i\xi\omega_c - \omega_c^2\right)\psi_c = 0, \tag{4.11a}$$

$$\psi_c(0) = 0, \quad \psi_c'(0) = 0, \tag{4.11b}$$

$$\psi_c''(1) + 2\mu_c\psi_c(1) = 0, \quad -\psi_c'''(1) - 2\mu_c\psi_c'(1) = 0. \tag{4.11c}$$

The (complex) left eigenvector ψ_c can be normalized as desired, for example with the aim to simplify future expressions (see the later Eq. 4.21).

Perturbation Solution

The solution of the generating perturbation Eq. 4.6 reads:

$$u_0 = A(t_2, \ldots)\phi_c(s)e^{i\omega_c t_0} + \text{c.c.}, \tag{4.12}$$

where a normalization condition, e.g., Re $(\phi_c(1)) = 1$ and Im $(\phi_c(1)) = 0$, is used to make the eigenvector unique[7]; moreover:

$$A = \frac{1}{2}a(t_2, \ldots)e^{i\varphi(t_2,\ldots)} \tag{4.13}$$

is an unknown complex amplitude, modulated on the slower time scales. When u_0 is substituted in the ϵ order perturbation Eq. 4.7, this latter becomes:

$$\partial_0^2 u_1 + \xi\partial_0 u_1 + u_1'''' + 2\mu_0 u_1'' + k_1 u_1 = -k_2(A^2\phi_c^2 e^{2i\omega_c t_0} + A\bar{A}\phi_c\bar{\phi}_c) + \text{c.c.}, \tag{4.14a}$$

$$u_1(0, t_k) = 0, \quad u_1'(0, t_k) = 0, \tag{4.14b}$$

$$u_1''(1, t_k) = 0, \quad -u_1'''(1, t_k) = 0. \tag{4.14c}$$

Since the known term is nonresonant (indeed, $e^{i\omega_c t}$ is missing in the right hand side of Eq. 4.14, and no internal resonances are assumed to occur), compatibility is satisfied. By neglecting the complementary solution, by virtue of the same arguments used in the previous chapter, the solution to Eqs. 4.14 is:

$$u_1 = A^2\chi_{12}(s)e^{2i\omega_c t_0} + A\bar{A}\chi_{10}(s) + \text{c.c.}, \tag{4.15}$$

where $\chi_{1m}(s)$, $m = 0, 2$, are solutions of the following problems[8]:

$$\chi_{1m}'''' + 2\mu_0\chi_{1m}'' + \left(k_1 + i\xi m\omega_c - m^2\omega_c^2\right)\chi_{1m} = \begin{cases} -k_2\phi_c^2(s), & \text{if } m=2, \\ -k_2\phi_c(s)\bar{\phi}_c(s), & \text{if } m=0, \end{cases} \tag{4.16a}$$

$$\chi_{1m}(0) = 0, \quad \chi_{1m}'(0) = 0, \tag{4.16b}$$

$$\chi_{1m}''(1) = 0, \quad -\chi_{1m}'''(1) = 0. \tag{4.16c}$$

With the previous results, the ϵ^2 order perturbation Eq. 4.8 reads:

$$\partial_0^2 u_2 + \xi\partial_0 u_2 + u_2'''' + 2\mu_0 u_2'' + k_1 u_2 = q_{21}(s)e^{i\omega_c t_0} + \text{N.R.T.} + \text{c.c.}, \tag{4.17a}$$

$$u_2(0, t_k) = 0, \quad u_2'(0, t_k) = 0, \tag{4.17b}$$

$$u_2''(1, t_k) = 0, \quad -u_2'''(1, t_k) = 0, \tag{4.17c}$$

---

[7] This is obtained by rescaling $\phi_c(s)$ as $\phi_c(s)/\phi_c(1)$, with $\phi_c(1)$ complex.

[8] Note that $\chi_{10}(s)$ is a real function.

where:

$$q_{21}(s) : = -(2i\omega_c + \xi)\partial_2 A\phi_c(s) - 2\tilde{\mu}A\phi_c''(s) - A^2\bar{A}\left[2k_2(2\phi_c(s)\chi_{10}(s) + \bar{\phi}(s)\chi_{12}(s)) + 3k_3\phi_c^2(s)\bar{\phi}_c(s)\right]. \quad (4.18)$$

The compatibility condition requires that the resonant terms to be orthogonal to the left eigenvector ψ_c, i.e.:

$$\int_0^1 \bar{\psi}_c(s)q_{21}(s)\,\mathrm{d}s = 0, \quad (4.19)$$

from which the unknown t_2 derivative of the amplitude is drawn:

$$\partial_2 A = c_1\tilde{\mu}A + c_3A^2\bar{A}, \quad (4.20)$$

where[9]:

$$c_1 := -\frac{2}{2i\omega_c + \xi}\frac{\int_0^1\bar{\psi}_c(s)\phi_c''(s)\mathrm{d}s}{\int_0^1\bar{\psi}_c(s)\phi_c(s)\mathrm{d}s}, \quad (4.21a)$$

$$c_3 := -\frac{2k_2}{2i\omega_c + \xi}\frac{2\int_0^1\bar{\psi}_c(s)\phi_c(s)\chi_{10}\mathrm{d}s + \int_0^1\bar{\psi}_c(s)\bar{\phi}_c(s)\chi_{12}\mathrm{d}s}{\int_0^1\bar{\psi}_c(s)\phi_c(s)\mathrm{d}s} - \frac{3k_3}{2i\omega_c + \xi}\frac{\int_0^1\bar{\psi}_c(s)\phi_c^2(s)\bar{\phi}_c(s)\mathrm{d}s}{\int_0^1\bar{\psi}_c(s)\phi_c(s)\mathrm{d}s}. \quad (4.21b)$$

To come back to the original variables, Eqs. 4.20 must be multiplied by ϵ^3, and the backward transformations $\epsilon^2\partial_2 A \to \dot{A}, \epsilon A \to A, \epsilon^2\tilde{\mu} \to \tilde{\mu}$ must be performed, leading to:

$$\dot{A} = c_1\tilde{\mu}A + c_3A^2\bar{A}. \quad (4.22)$$

This is called the *bifurcation equation* of the problem. It governs the dynamic of a reduced two-dimensional dynamical system, asymptotically equivalent to the original infinite-dimensional system.

Equation 4.22 is recast in the real form by making use of the polar representation for the amplitude (Eq. 4.13), i.e.:

$$\dot{a} = c_{1R}(\mu - \mu_c)\,a + \frac{1}{4}c_{3R}a^3, \quad (4.23a)$$

$$a\dot{\varphi} = c_{1I}(\mu - \mu_c)\,a + \frac{1}{4}c_{3I}a^3, \quad (4.23b)$$

[9] Equations 4.21 could be simplified if we normalized the left eigenvector according to $\int_0^1\bar{\psi}_c(s)\phi_c(s)\,\mathrm{d}s = 1$.

where $c_{jR} := \mathrm{Re}(c_j), c_{jI} := \mathrm{Im}(c_j)$, with $j = 1, 3$, and $\tilde{\mu} = \mu - \mu_c$. An important feature of these equations is that they can be solved in cascade, since they are of the type $\dot{a} = f(a)$, $\dot{\varphi} = g(a)$. Once they have been integrated, possibly numerically, the response of the beam follows from Eqs. 4.4, 4.12 and 4.15, with ϵ reabsorbed; it reads:

$$\begin{aligned} u(s,t) = {} & a(t)[\mathrm{Re}(\phi_c(s))\cos(\omega_c t + \varphi(t)) - \mathrm{Im}(\phi_c(s))\sin(\omega_c t + \varphi(t))] + \frac{a^2}{2}[\chi_{10}(s) \\ & + \mathrm{Re}(\chi_{12}(s))\cos(2(\omega_c t + \varphi(t))) - \mathrm{Im}(\chi_{12}(s))\sin(2(\omega_c t + \varphi(t)))]. \end{aligned} \tag{4.24}$$

Remark 4.3 The complex bifurcation Eq. 4.22, or, equivalently, its real counterpart Eqs. 4.23, encompasses the essence of the *Center Manifold Theorem* [1, 2, 4, 5]. It affirms that the long-term dynamics of a system close to a bifurcation takes place on an *invariant* manifold of the state space,[10] whose dimension equates that of the critical subspace, and which is tangent at the origin to the linear subspace spanned by the critical eigenvectors. In the Hopf bifurcation case, we have two critical eigenvalues $\lambda_c = \pm i\omega_c$, and a two-dimensional subspace, spanned by $\mathrm{Re}(\phi_c), \mathrm{Im}(\phi_c)$; therefore the motion occurs on a two-dimensional manifold, tangent to this plane, according to Eq. 4.24. The meaning of this fundamental theorem is that nonlinearities *deform* the critical space, but they do not change its dimension.

Remark 4.4 As we already noticed in the past problems, quadratic nonlinearities trigger the passive (i.e., stable!) modes, whose contributions is described by the functions $\chi_{1m}(s)$. Therefore, stable modes cannot be neglected in a (nonlinear) bifurcation analysis. However, they do not participate to the motion with their own free dynamics (as it is evident from the fact that $\chi_{1m}(s)$ are particular solutions to forcing excitations), but they only contribute to the response as modes driven by the (unique unstable) active mode. Since this mode is complex, i.e., the linear motion is a superposition of two shapes ($\mathrm{Re}(\phi_c(s)), \mathrm{Im}(\phi_c(s))$ oscillating in quadrature), the equivalent dynamical system is also two-dimensional. These latter are the underlying ideas on which the Center Manifold Theory is grounded, which permits reduction of the system to a small-dimensional system.

Limit Cycles

A steady motion of the beam occurs when $a = a_e =$ const. From Eq. 4.23a two solutions are found for a_e[11]:

---

[10] 'Invariant' manifold means that if the initial conditions belong to it, the motion does not leave it.

[11] The negative value of the square root is discarded, since negative amplitudes simply repeat positive amplitudes opposite in phase.

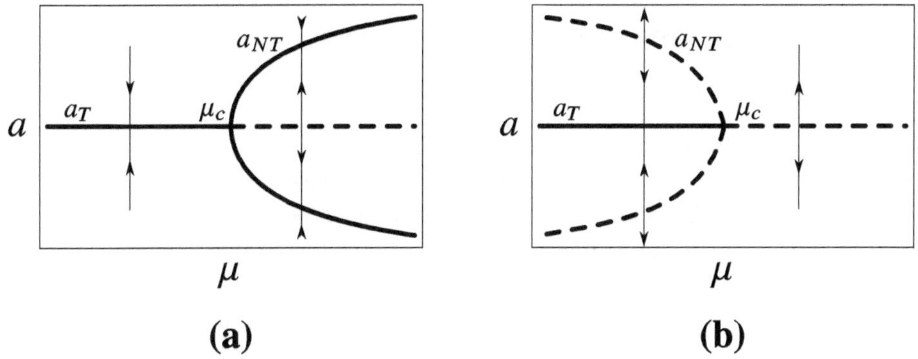

Fig. 4.1 Bifurcation diagram for the Hopf bifurcation: **a** super-critical, **b** sub-critical; a_T trivial solution, a_{NT} non-trivial solution. Continuous lines: stable, dashed lines: unstable. Arrows denote the evolution of the amplitude

$$a_e = \begin{cases} 0 & =: a_T, \\ 2\sqrt{-\frac{c_{1R}}{c_{3R}}(\mu - \mu_c)} & =: a_{NT}, \end{cases} \tag{4.25}$$

which will be referred to as *trivial* and *non-trivial* solution, respectively. The steady solutions are plotted in the bifurcation diagram of Fig. 4.1.

It should be noticed, that the nontrivial solution only exists on one side of the bifurcation diagram, according to the sign of c_{1R}/c_{3R}. If this ratio is negative, then the domain of existence is $\mu > \mu_c$, and the bifurcation is said *super-critical* (see Fig. 4.1a); if this ratio is positive, the domain of existence is $\mu < \mu_c$, and the bifurcation is said *sub-critical* (see Fig. 4.1b). The bifurcation diagram shows that, when the bifurcation is super-critical, the amplitude of oscillation does not diverge to infinity in the postcritical range, as predicted by the linear theory, but it is confined by nonlinearities to an amplitude, a_{NT}, which depends on the intensity of the tangential force.

By substituting Eq. 4.25 in Eq. 4.23b and integrating, it follows that:

$$\varphi_e = \begin{cases} \varphi_0, & \text{if } a_e = 0, \\ \Delta\omega t + \varphi_0, & \text{if } a_e \neq 0, \end{cases} \tag{4.26}$$

where:

$$\Delta\omega := c_{1I}(\mu - \mu_c) + \frac{1}{4} c_{3I} a_e^2 \tag{4.27}$$

is the *frequency correction* and φ_0 an initial phase.

By summarizing, the steady motion is periodic, since the amplitude is constant and the phase-modulation is linear, this resulting in a modification of the linear frequency of the response in Eq. 4.24; the motion is expressed by:

$$u = a_e[\mathrm{Re}(\phi_c(s))\cos(\varpi t + \varphi_0) - \mathrm{Im}(\phi_c(s))\sin(\varpi t + \varphi_0)] + \frac{a_e^2}{2}[\chi_{10}(s) \\ + \mathrm{Re}(\chi_{12}(s))\cos(2(\varpi t + \varphi_0)) - \mathrm{Im}(\chi_{12}(s))\sin(2(\varpi t + \varphi_0))], \quad (4.28)$$

where $\varpi := \omega_c + \Delta\omega$ is the (amplitude-dependent) *nonlinear frequency*.

To detect (orbital) stability of the periodic motion, we have to analyze the stability of the equilibrium point a_e. Therefore, we build-up the variational equation of Eq. 4.23a (only), by letting $a = a_e + \delta a$ and linearizing it around the perturbation $\delta a = 0$, thus obtaining:

$$\delta\dot{a} = \left[c_{1R}(\mu - \mu_c) + \frac{3}{4}c_{3R}a_e^2\right]\delta a. \quad (4.29)$$

This equation must be evaluated separately on the trivial and non-trivial solutions, in order to check stability of each of them. By substituting Eq. 4.25, it becomes:

$$\delta\dot{a} = \delta a \begin{cases} c_{1R}(\mu - \mu_c), & \text{if } a_e = 0, \\ -2c_{1R}(\mu - \mu_c), & \text{if } a_e \neq 0. \end{cases} \quad (4.30)$$

The equilibrium is stable when the coefficient on the right hand side of Eq. 4.30 is negative (since δa decays exponentially), and unstable when it is positive (since δa increases exponentially). It is found that at the bifurcation point $\mu = \mu_c$ the two branches change stability. Since the trivial branch is stable when $\mu < \mu_c$ and unstable when $\mu > \mu_c$, then $c_{1R} > 0$. Therefore the non-trivial solution is stable when $\mu > \mu_c$ (i.e., when the bifurcation is super-critical), and unstable when $\mu < \mu_c$ (i.e., when the bifurcation is sub-critical).

Periodic motions are represented by circular orbits in the polar phase-plane (a, φ), said *limit cycles*. A stable limit cycle attracts all the surrounding orbits (Fig. 4.2a), this entailing that the beam experiences a transient motion until it reaches a periodic regime. An unstable limit cycle (Fig. 4.2b) repels the surrounding orbits, so that the beam either comes back to

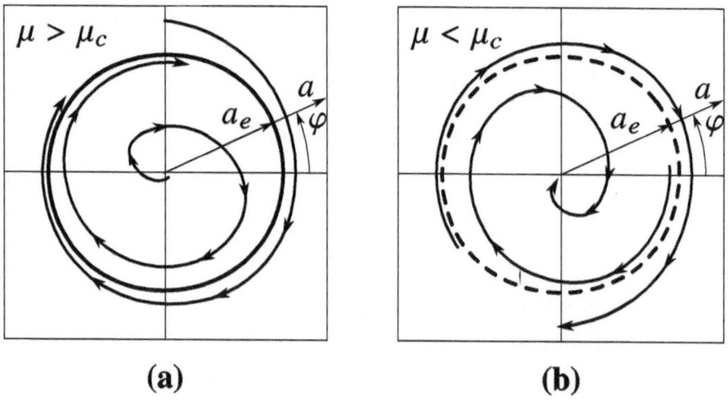

Fig. 4.2 Limit cycles and surrounding orbits at a Hopf bifurcation point, in polar coordinates: **a** super-critical, **b** sub-critical bifurcations

the rest position (if the initial perturbation is small), or it moves away somewhere else, if the perturbation is large.[12] This last occurrence is very dangerous, and it likely causes the failure of the beam.

Remark 4.5 The bifurcation diagram of Fig. 4.1 resembles the pitchfork we have already encountered in buckling of symmetric systems (remember Fig. 2.2). However, it should be kept in mind, that now we are dealing with the amplitude of the response, not with the response itself. In other words, we are studying the motion after *the fast dynamics has been filtered*, in such a way a periodic response appears as an equilibrium point for the amplitude equation.

4.2.2 Symmetric Systems

When $k_2 = 0$, we can, as usual, skip the trivial ϵ order Eq. 4.7, by omitting the odd-terms of the asymptotic series and renaming by ϵ the old squared perturbation parameter. Thus, we rescale the response as $u \to \epsilon^{1/2}\hat{u}$; we still split the bifurcation parameter as $\mu = \mu_0 + \tilde{\mu}$, but we use the ordering $\tilde{\mu} \to \epsilon\tilde{\mu}$; finally we expand the response as $u(s, t_0, t_1, \ldots) = u_0(s, t_0, t_1, \ldots) + \epsilon u_1(s, t_0, t_1, \ldots) + \cdots$, and the time differential operators as $\partial_t = \partial_0 + \epsilon\partial_1 + \cdots$ and $\partial_t^2 = \partial_0^2 + 2\epsilon\partial_0\partial_1 + \cdots$. Accordingly, we obtain the following perturbation equations:

Order ϵ^0:

$$\partial_0^2 u_0 + \xi\,\partial_0 u_0 + u_0'''' + 2\mu_0 u_0'' + k_1 u_0 = 0, \tag{4.31a}$$

$$u_0(0, t_k) = 0, \quad u_0'(0, t_k) = 0, \tag{4.31b}$$

$$u_0''(1, t_k) = 0, \quad -u_0'''(1, t_k) = 0. \tag{4.31c}$$

Order ϵ^1:

$$\partial_0^2 u_1 + \xi\,\partial_0 u_1 + u_1'''' + 2\mu_0 u_1'' + k_1 u_1 = -2\partial_0\partial_1 u_0 - \xi\,\partial_1 u_0 - 2\tilde{\mu} u_0'' - k_3 u_0^3, \tag{4.32a}$$

$$u_1(0, t_k) = 0, \quad u_1'(0, t_k) = 0, \tag{4.32b}$$

$$u_1''(1, t_k) = 0, \quad -u_1'''(1, t_k) = 0. \tag{4.32c}$$

Equation 4.31 is identical to Eq. 4.6. By replacing its generating solution, Eq. 4.12, in Eq. 4.32, this latter becomes:

[12] We cannot predict where the beam goes, since the perturbation analysis is of *local* type.

$$\partial_0^2 u_1 + \xi \partial_0 u_1 + u_1'''' + 2\mu_0 u_1'' + k_1 u_1 = q_{11}(s) e^{i\omega_r t_0} + \text{N.R.T.} + \text{c.c.}, \tag{4.33a}$$

$$u_1(0, t_k) = 0, \quad u_1'(0, t_k) = 0, \tag{4.33b}$$

$$u_1''(1, t_k) = 0, \quad -u_1'''(1, t_k) = 0, \tag{4.33c}$$

where:

$$q_{11}(s) := -(2i\omega_c + \xi)\partial_1 A\phi_c(s) - 2\tilde{\mu} A\phi_c''(s) - 3k_3 A^2 \bar{A} \phi_c^2(s)\bar{\phi}_c(s). \tag{4.34}$$

Enforcing compatibility, $\partial_1 A$ is evaluated, and coming back to the true variables, the bifurcation Eq. 4.22 is recovered, with c_1 still given by Eq. 4.21a, and c_3 updated as follows:

$$c_3 = -\frac{3k_3}{2i\omega_c + \xi} \frac{\int_0^1 \bar{\psi}_c(s)\phi_c^2(s)\bar{\phi}_c(s)\,\mathrm{d}s}{\int_0^1 \bar{\psi}_c(s)\phi_c(s)\,\mathrm{d}s}. \tag{4.35}$$

The response of the system, at this order, remains the linear one, but with amplitude and phases modulated according to Eqs. 4.23, i.e.:

$$u = a(t)[\mathrm{Re}(\phi_c(s))\cos(\omega_c t + \varphi(t)) - \mathrm{Im}(\phi_c(s))\sin(\omega_c t + \varphi(t))]. \tag{4.36}$$

Remark 4.6 As we already observed, the lack of quadratic nonlinearities entails the absence of passive contributions in the first asymptotic approximation.

Remark 4.7 The Hopf bifurcation is *robust*, since imperfections cannot destroy it. Indeed, a known constant term, as a force αP_B, cannot enter in the bifurcation equation, since it is nonresonant. In contrast, a perturbation of the stiffness, e.g., a modifications of the soil coefficient $\delta k_1(s)$, generates a perturbation $\delta k_1(s)\, u$ which is resonant. However, this does not introduce qualitative new terms in the bifurcation equations, thus leaving the behavior of the imperfect system substantially unaltered, with respect to the perfect system.

4.3 Static Bifurcations of Dynamical Systems

In Sect. 2.3 we tackled the buckling problem for a conservative static system. By applying the strained parameter method, we were able to find the equilibrium paths bifurcating from the fundamental path, but we were unable to check their stability, and postponed the relevant discussion. Now, we want to address the problem into the larger context of *static bifurcations of dynamical systems* (also said *divergence bifurcation*).

Two-Parameter Beam

When an elastic system is subject to conservative forces, it can lose stability only by divergence; in contrast, when the forces are nonconservative, loss of stability can manifest itself either by a Hopf bifurcation or divergence.[13] We have already analyzed dynamical bifurcation in Sect 4.2; here we want to analyze divergence. To this end, we consider a beam on elastic soil, prestressed by *both* conservative $\nu_s = \nu$ and nonconservative μ forces, and we want to investigate the mechanical behavior of the system around a divergence point. The equation of motion ruling the dynamics of the two-parameter model is Eq. 1.7, made homogeneous and with $\nu_d = 0$, i.e.[14]:

$$\ddot{u} + \xi \dot{u} + u'''' + 2(\mu + \nu) u'' + k_1 u + k_2 u^2 + k_3 u^3 + \cdots = 0, \tag{4.37a}$$

$$u(0, t) = 0, \quad u'(0, t) = 0, \tag{4.37b}$$

$$u''(1, t) = 0, \quad -u'''(1, t) - 2\nu u'(1, t) = 0. \tag{4.37c}$$

Of course, one can ask if such a bifurcation can exist in presence of both forces. It is easy to realize that this is, indeed, possible. In fact, we know that, when the beam is prestressed by the dead load only, it undergoes bifurcation at the critical (or Eulerian) load ν_{c_k}. If we add a sufficiently small nonconservative force, and we invoke continuity of the solution with parameters, such a bifurcation persists. From a geometrical point of view, if there exist a point $E_k := (0, \nu_{c_k})$ on the (μ, ν)-plane, which is an Eulerian bifurcation point, then it cannot be isolated, but it must belong to some branch $\mathcal{D}$ that extends itself into the neighborhood of the point (see Fig. 4.3[15]).

Multiple Scale Method

We will tackle Eq. 4.37 by the Multiple Scale Method, pursuing a *first order solution*. Accordingly, we rescale the variable as $u \rightarrow \epsilon \hat{u}$; then we split the *two* bifurcation parameters as:

$$\mu = \mu_0 + \tilde{\mu}, \qquad \nu = \nu_0 + \tilde{\nu} \tag{4.38}$$

[13] However, the system analyzed in Sect. 4.2 does not exhibit divergence, but only dynamic bifurcation. Therefore, to make it possible, a gravitational force has to be added to the nonconservative system.

[14] A discrete version of the problem is dealt with in the Sect. B.7 of the Appendix B.

[15] Here, differently from the dynamic case, both positive and negative amplitudes are of interest, since the bifurcation is static and no phase exists.

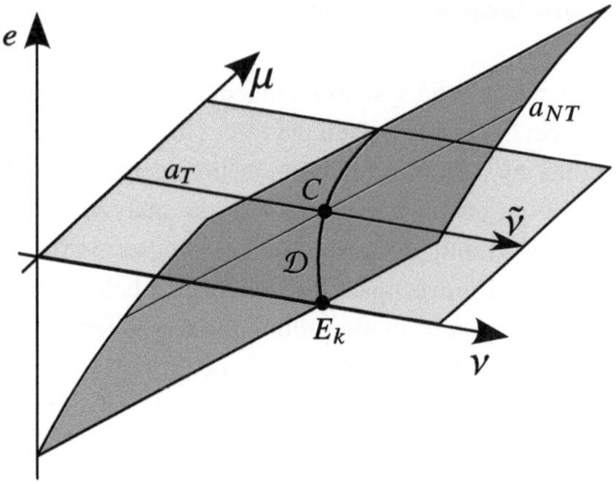

Fig. 4.3 Beam under conservative ν and nonconservative μ prestresses. Divergence locus $\mathcal{D}$ in the (μ, ν)-parameter plane and equilibrium surfaces; Eulerian critical point $E_k := (0, \nu_{ck})$; generic critical point $C := (\mu_c, \nu_c)$; trivial $a_T(\tilde{\nu})$ and non-trivial $a_{NT}(\tilde{\nu})$ equilibrium paths at $\mu = \mu_c$

and order their incremental parts as $\tilde{\mu} \to \epsilon\tilde{\mu}$, $\tilde{\nu} \to \epsilon\tilde{\nu}$.[16] As in the previous Section, to avoid coalescence of eigenvalues, we do *not* rescale the damping, but assume it is of order 1.[17] Finally, we expand the variable as:

$$u(s, t_0, t_1, \ldots) = u_0(s, t_0, t_1, \ldots) + \epsilon u_1(s, t_0, t_1, \ldots) + \cdots \tag{4.39}$$

and the differential operators as $\partial_t = \partial_0 + \epsilon\partial_1 + \cdots$, $\partial_t^2 = \partial_0^2 + 2\epsilon\partial_0\partial_1 + \cdots$. The following perturbation equations are drawn:

Order ϵ^0:

$$\partial_0^2 u_0 + \xi\partial_0 u_0 + u_0'''' + 2(\mu_0 + \nu_0)u_0'' + k_1 u_0 = 0, \tag{4.40a}$$

$$u_0(0, t_k) = 0, \quad u_0'(0, t_k) = 0, \tag{4.40b}$$

$$u_0''(1, t_k) = 0, \quad -u_0'''(1, t_k) - 2\nu_0 u_0'(1, t_k) = 0. \tag{4.40c}$$

[16] Remember that, by studying the Hopf bifurcation, we rescaled the increment of the parameter at the ϵ^2 order, since quadratic nonlinearities were unable to produce ω_c resonant terms at the ϵ order. Here, instead, the quadratic terms will produce resonant terms, as it will appear clear soon.

[17] To better understand the problem, if we consider a single d.o.f. system $\ddot{x} + \xi\dot{x} + \nu x = 0$, the relevant characteristic equation reads $\lambda^2 + \xi\lambda + \nu = 0$. The system undergoes a divergence ($\lambda = 0$) when $\nu = 0$, if and only if $\xi \neq 0$. If, in contrast, $\xi = 0$, then a double-zero ($\lambda = 0, 0$) bifurcation takes place at $\nu = 0$, which is a much more difficult problem to be studied.

Order ϵ^1:

$$\begin{aligned}\partial_0^2 u_1 + \xi \partial_0 u_1 + u_1'''' + 2(\mu_0 + \nu_0) u_1'' + k_1 u_1 &= -2\partial_0\partial_1 u_0 - \xi \partial_1 u_0 \\ &\quad - 2(\tilde{\mu} + \tilde{\nu}) u_0'' - k_2 u_0^2, \end{aligned} \tag{4.41a}$$

$$u_1(0, t_k) = 0, \quad u_1'(0, t_k) = 0, \quad u_1''(1, t_k) = 0, \tag{4.41b}$$

$$- u_1'''(1, t_k) - 2\nu_0 u_1'(1, t_k) = 2\tilde{\nu} u_0'(1, t_k). \tag{4.41c}$$

Right and Left Eigenvectors

The generating Eq. 4.40 is studied first. By letting $u_0 = \phi_k(s) e^{\lambda_k t_0}$, a space boundary value problem follows:

$$\phi_k'''' + 2(\mu_0 + \nu_0)\phi_k'' + \left(k_1 + \xi\lambda_k + \lambda_k^2\right)\phi_k = 0, \tag{4.42a}$$

$$\phi_k(0) = 0, \quad \phi_k'(0) = 0, \tag{4.42b}$$

$$\phi_k''(1) = 0, \quad -\phi_k'''(1) - 2\nu_0\phi_k'(1) = 0. \tag{4.42c}$$

As we discussed above, there exist infinite pairs (μ_0, ν_0) at which a simple eigenvalue $\lambda_0 := \lambda(\mu_0, \nu_0) = 0$ vanishes (with the remaining eigenvalues having negative real part). They describe a divergence locus $\mathcal{D}$ on the parameter plane (see Fig. 4.3). We take *one* of these points, denoted by $C := (\mu_c, \nu_c)$, and we focus our attention on its neighborhood in the plane. Therefore $(\lambda_c = 0, \phi_c)$ is an eigenpair of Eq. 4.42, when $(\mu_0, \nu_0) = (\mu_c, \nu_c)$, i.e.:

$$\phi_c'''' + 2(\mu_c + \nu_c)\phi_c'' + k_1\phi_c = 0, \tag{4.43a}$$

$$\phi_c(0) = 0, \quad \phi_c'(0) = 0, \tag{4.43b}$$

$$\phi_c''(1) = 0, \quad -\phi_c'''(1) - 2\nu_c\phi_c'(1) = 0. \tag{4.43c}$$

The (right) eigenvector ϕ_c is real, and we will assume it has been suitably normalized, e.g., by $\phi_c(1) = 1$. The previous equations are the homogeneous version of Eqs. A.13 studied in the Appendix A, so that the adjoint problem follows from Eqs. A.15:

$$\psi_c'''' + 2(\mu_c + \nu_c)\psi_c'' + k_1\psi_c = 0, \tag{4.44a}$$

$$\psi_c(0) = 0, \quad \psi_c'(0) = 0, \tag{4.44b}$$

$$\psi_c''(1) + 2\mu_c\psi_c(1) = 0, \quad -\psi_c'''(1) - 2(\mu_c + \nu_c)\psi_c'(1) = 0. \tag{4.44c}$$

From this latter problem, the (real) left eigenvector ψ_c is evaluated, to be normalized as desired. Since the problem is not self-adjoint, $\psi_c \neq \phi_c$; if, as a special case, $\mu_c = 0$, then $\psi_c \equiv \phi_c$.

Perturbation Solution

According to the previous discussion, the generating solution of the Eqs. 4.40 reads:

$$u_0 = a(t_1, t_2, \ldots)\phi_c(s), \qquad (\mu_0, \nu_0) = (\mu_c, \nu_c), \tag{4.45}$$

where $a(t_1, t_2, \ldots)$ is a *real* amplitude, slowly modulated in time, which represents the response at the free end of the beam.[18] When this solution is substituted in Eq. 4.41, this latter becomes:

$$\begin{aligned} \partial_0^2 u_1 + \xi \partial_0 u_1 + u_1'''' + 2(\mu_c + \nu_c)u_1'' + k_1 u_1 &= -\xi \partial_1 a\phi_c(s) \qquad (4.46\text{a})\\ &\quad - 2(\tilde{\mu} + \tilde{\nu})a\phi_c''(s) - k_2 a^2 \phi_c^2(s), \\ u_1(0, t_k) = 0, \quad u_1'(0, t_k) = 0, \quad u_1''(1, t_k) &= 0, \qquad (4.46\text{b})\\ - u_1'''(1, t_k) - 2\nu_c u_1'(1, t_k) &= 2\tilde{\nu}a\phi_c'(1). \qquad (4.46\text{c}) \end{aligned}$$

This equations are similar to Eqs. A.19, but with a forcing frequency independent of t_0, namely $p(s, t_0) = q(s)$, $P_B(t_0) = Q_B$. In the terminology of harmonic analysis, we can also say that the forcing term has frequency $\Omega = 0$. However, since the associated homogeneous problem admits the 0-frequency ($\lambda_c = 0$) as an eigenvalue, the know term is 'resonant', and therefore it has to satisfy the compatibility condition of Eq. A.23, i.e.:

$$-\int_0^1 \psi_c(s)[\xi \partial_1 a\phi_c(s) + 2(\tilde{\mu} + \tilde{\nu})a\phi_c''(s) + k_2 a^2 \phi_c^2(s)]\,\mathrm{d}s + 2\tilde{\nu}a\psi_c(1)\phi_c'(1) = 0. \tag{4.47}$$

From this equations, the unknown $\partial_1 a$ is derived:

$$\partial_1 a = \frac{1}{\xi}\left[(c_{11}\tilde{\mu} + c_{12}\tilde{\nu})a + c_{20}a^2\right], \tag{4.48}$$

where, after an integration by parts involving c_{12}, it is:

[18] Compare this solution with Eq. 2.23 relevant to static systems. The only substantial novelty stands in the fact that here the amplitude is *not* a constant but it is a function of the slow time.

$$c_{11} = -\frac{2\int_0^1 \psi_c(s)\phi_c''(s)\,\mathrm{d}s}{\int_0^1 \psi_c(s)\phi_c(s)\,\mathrm{d}s}, \qquad c_{12} = \frac{2\int_0^1 \psi_c'(s)\phi_c'(s)\,\mathrm{d}s}{\int_0^1 \psi_c(s)\phi_c(s)\,\mathrm{d}s},$$
$$c_{20} = -\frac{k_2\int_0^1 \psi_c(s)\phi_c^2(s)\,\mathrm{d}s}{\int_0^1 \psi_c(s)\phi_c(s)\,\mathrm{d}s}. \tag{4.49}$$

If the analysis is truncated here, $\dot{a} = \epsilon\partial_1 a$ must be considered, which leads, after considering the backward rescaling $\epsilon a \to a$ and $\epsilon(\tilde{\mu}, \tilde{\nu}) \to (\tilde{\mu}, \tilde{\nu})$, to the *first order bifurcation equation*:

$$\dot{a} = \frac{1}{\xi}\left[(c_{11}\tilde{\mu} + c_{12}\tilde{\nu})a + c_{20}a^2\right]. \tag{4.50}$$

Equation 4.50 governs the amplitude modulation, and constitutes an equivalent one-dimensional system.[19] Its dynamics depend on two parameters, $\tilde{\mu}$ and $\tilde{\nu}$, which permit to explore a ball around the critical point C, in the (μ, ν)-plane. However, sufficient *qualitative* information can be drawn by exploring a segment *transverse* to the divergence locus, e.g., by using just one of them, or a linear combination of them.[20, 21] For example, if we choose $\tilde{\nu}$ as preferred bifurcation parameter and we put $\tilde{\mu} = 0$ (see Fig. 4.3), the bifurcation equation simplifies as follows:

$$\dot{a} = \frac{1}{\xi}\left(c_{12}a\tilde{\nu} + c_{20}a^2\right). \tag{4.51}$$

If we look for equilibrium points $a = a_e = \text{const}$, we have to consider $\dot{a} = 0$ in Eq. 4.51 and to solve an algebraic equation. One root is trivial, $a_e = 0 =: a_T$, the other is non-trivial, $a_e = a_{NT} = -\frac{c_{12}\tilde{\nu}}{c_{20}}$. To analyze stability, we perform the variation of the bifurcation Eq. 4.51, thus getting:

$$\delta\dot{a} = \frac{1}{\xi}\left(c_{12}\tilde{\nu} + 2c_{20}a_e\right)\delta a. \tag{4.52}$$

By substituting $a_e = 0$, and remembering that the trivial solution loses stability when $\tilde{\nu} > 0$, it follows that $c_{12} > 0$. By substituting the non-trivial solution, the variational equation becomes:

[19] This agrees with the Center Manifold Theory, since divergence entails just one zero-eigenvalue, so that the critical subspace is one-dimensional, spanned by the real eigenvector ϕ_c.

[20] In this way we are looking for an equilibrium branch which is a cross-section of a surface of equilibrium points (see Fig. 4.3).

[21] This is a fundamental rule, well-known in Bifurcation Theory, where it is stated that the number of parameters *strictly* necessary to describe a bifurcation is equal to the *codimension* of the bifurcation. Divergence (as well as Hopf) bifurcations are codimension-1 bifurcations, since a unique degenerateness condition is requested to the eigenvalues of the linearized system, i.e., $\lambda = 0$ (or $\mathrm{Re}(\lambda) = 0$). Therefore, bifurcation loci are points in 1-dimensional parameter-spaces, or lines in two-dimensional spaces, or surfaces in three-dimensional spaces, and so on. A transverse locus to these loci, irrespective of the dimension of the system, is always a line, i.e., a one-dimensional manifold, on which the 'postcritical' behavior is described by just one parameter.

$$\delta\dot{a} = -\frac{1}{\xi}c_{12}\tilde{\nu}\delta a, \tag{4.53}$$

from which it follows that, close to $\tilde{\nu} = 0$, super-critical branches are stable, and sub-critical unstable, according to the results anticipated in Sect. 2.3.

Remark 4.8 The MSM furnishes the same equilibria one could (more easily) find by the strained parameter method. The reader can verify that the latter method also works in the problem at hand, i.e., in presence of nonconservative forces. Therefore, one could ask 'why to apply the MSM?'. The response is in the 'dynamic' part of the bifurcation Eq. 4.51, which permits us to investigate stability, too. As already noticed, it is true that stability could be checked by using the Total Potential Energy criterion, but this works only for conservative systems. For nonconservative systems, indeed, a dynamic analysis, as that carried out here, is the unique way to ascertain stability.

4.4 Exercises

The MSM is applied to solve some case studies relevant to the beam on elastic soil, namely: (i) the dynamic bifurcation, occurring when the beam is compressed by a tangential force; (ii) the static bifurcation, manifesting when both a tangential and a gravitational compression forces act on the beam.

Exercise 4.1 (*Dynamic bifurcation of a beam on elastic soil*) Consider a beam on non-symmetric Winkler soil, prestressed by a tangential tip force μ. (a) Plot the constitutive law of the soil. (b) Determine the critical load and frequency and plot the critical mode. (c) Represent the passive displacement fields $\chi_{10}(s)$, $\chi_{12}(s)$ and plot the bifurcation diagram. (d) Determine the steady time evolution of the tip of the beam in postcritical conditions. Take the following numerical values: $k_1 = 50$, $k_2 = 100$, $k_3 = 300$, $\xi = 0.1$.

(a) The soil constitutive law, $f(u) = k_1 u + k_2 u^2 + k_3 u^3$, is plotted in Fig. 4.4a, where (I) linear, (II) linear plus quadratic, and (III) linear plus quadratic and cubic contributions are shown, highlighting the magnitude of nonlinearities in the range of displacements explored.

(b) The critical values μ_c and ω_c must be sought tackling the eigenvalue problem Eqs. 4.10. The relevant characteristic equation reads:

$$\lambda^4 + 2\mu_c\lambda^2 + (k_1 + i\xi\omega_c - \omega_c^2) = 0, \tag{4.54}$$

whose (complex) roots are:

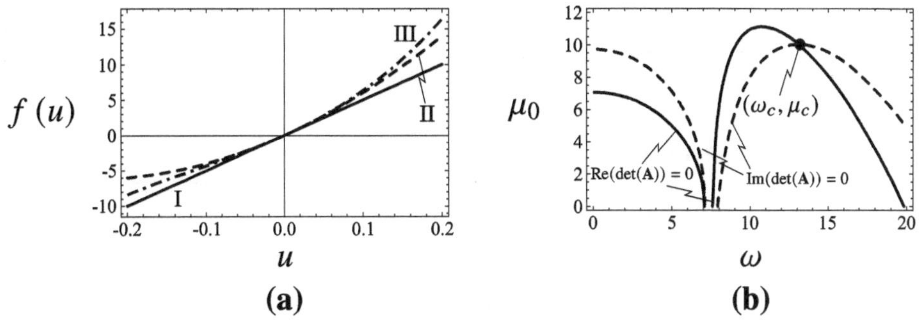

Fig. 4.4 **a** Constitutive law of the soil, highlighting (I) linear, (II) linear plus quadratic, and (III) linear plus quadratic and cubic contributions; **b** evaluation of the critical condition (ω_c, μ_c)

$$\lambda_{1,2} = \pm\sqrt{-\mu_c + \sqrt{\mu_c^2 + \omega_c^2 - i\xi\omega_c - k_1}}, \tag{4.55a}$$

$$\lambda_{3,4} = \pm\sqrt{-\mu_c - \sqrt{\mu_c^2 + \omega_c^2 - i\xi\omega_c - k_1}}. \tag{4.55b}$$

Accordingly, the general solution of Eq. 4.10a reads:

$$\phi_c(s) = \sum_{j=1}^{4} a_j e^{\lambda_j s}. \tag{4.56}$$

By substituting it in the boundary conditions Eqs. 4.10b, c, a homogeneous algebraic system $\mathbf{Aa} = \mathbf{0}$ follows, whose matrix $\mathbf{A}$ should be made singular. Concurrently vanishing real and imaginary parts of the determinant of $\mathbf{A}$, provides couples (μ_c, ω_c). The Fig. 4.4b shows a graphical constructions in the (μ, ω) plane, which shows the pair of roots with the lowest μ_c, namely $(\mu_c, \omega_c) = (10.03,\ 13.09)$.

For such a pair (μ_c, ω_c), the critical mode $\phi_c(s)$ is obtained by solving Eqs. 4.10. The same critical values allow one to solve the adjoint problem Eqs. 4.11, to get $\psi_c(s)$. Their real and imaginary parts are shown in Fig. 4.5a, b, respectively.

(c) The passive displacement fields $\chi_{10}(s)$ and $\chi_{12}(s)$ are evaluated by solving Eqs. 4.16. The real and imaginary parts of the χ functions ($\chi_{10}(s)$ is real) are shown in Fig. 4.5c, d, respectively. With these results, the coefficients c_1 and c_3 are evaluated from Eqs. 4.21, to get: $c_1 = 66.993 - 0.245i$, $c_3 = -396.962 + 7.374i$. The bifurcation diagram supplied by Eq. 4.25 is plotted in Fig. 4.6; a super-critical Hopf bifurcation is found.

(d) The steady time evolution of $u(1, t)$ is drawn from Eq. 4.28; when $\mu = 10.1$, it is shown in Fig. 4.6b, where (I) linear and (II) linear plus quadratic solutions are distinguished, in continuous and dashed lines, respectively. It is seen that higher order term brings a small contribution to the motion. □

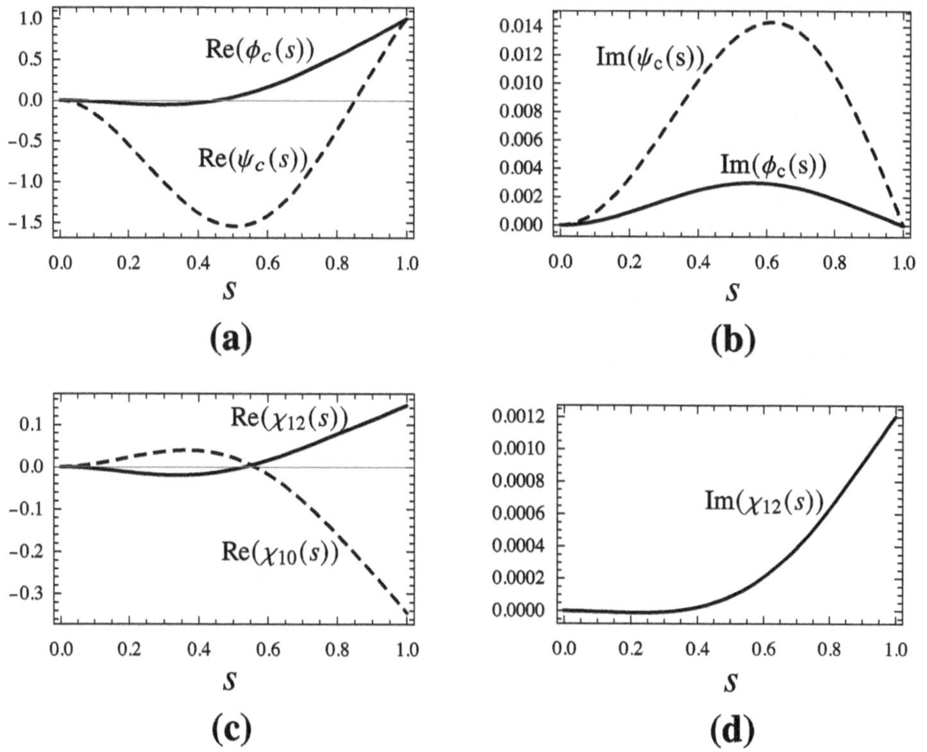

Fig. 4.5 Right eigenvector $\phi_c(s)$ of Eq. 4.10 and left eigenvector $\psi_c(s)$ of Eqs. 4.11: **a** real part, **b** imaginary part; passive displacement fields $\chi_{10}(s)$ and $\chi_{12}(s)$: **c** real part, **d** imaginary part

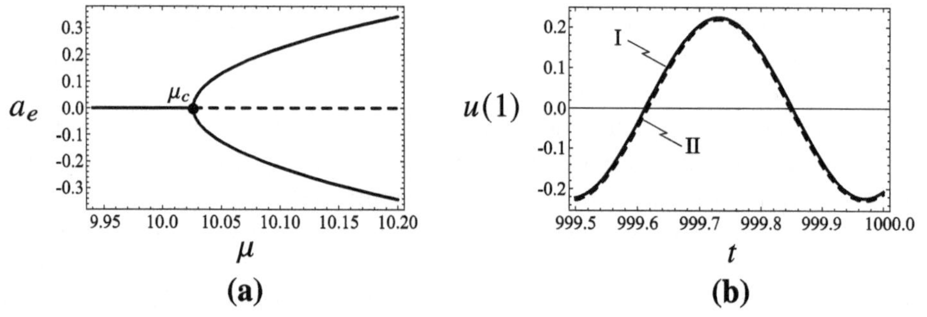

Fig. 4.6 Hopf bifurcation of beam on elastic soil, when $k_1 = 50$, $k_2 = 100$, $k_3 = 300$: **a** bifurcation diagram (continuous line: stable; dashed line: unstable), **b** displacement of the tip of the beam when $\mu = 10.2$: (I) ϵ^0 order solution (continuous line), (II) ϵ^1 order solution (dashed line)

Exercise 4.2 (*Static bifurcation of a beam on elastic soil*) Let us consider a beam on non-symmetric Winkler soil, prestressed by both a dead ν *and a follower* μ load. (a) Determine the divergence locus on the (μ, ν) plane. (b) Take $\mu_c = 0$ on this locus, and evaluate: (i) the critical right $\phi_c(s)$ and left $\psi_c(s)$ eigenvectors, (ii) the bifurcation diagram, depicting the displacement of tip of the beam vs the bifurcation parameter ν. (c) Repeat the previous analysis by taking $\mu_c = 2$. Make use of the following values of the mechanical characteristics: $k_1 = 1, k_2 = 100$.

(a) The critical values μ_c and ν_c are supplied by the eigenvalue problem in Eqs. 4.43, whose characteristic equation reads:

$$\lambda^4 + 2(\mu_c + \nu_c)\lambda^2 + k_1 = 0. \tag{4.57}$$

Under the hypothesis $(\mu_c + \nu_c)^2 > k_1$ (to be verified a posteriori[22]), it admits the following roots: $\lambda_{1,2} := \pm i\beta_1, \lambda_{3,4} := \pm i\beta_2$, where:

$$\beta_{1,2} := \sqrt{\mu_c + \nu_c \pm \sqrt{(\mu_c + \nu_c)^2 - k_1}}, \tag{4.58}$$

which are real values. Therefore, the right eigenvector reads:

$$\phi_c(s) = a_1 \cos(\beta_1 s) + a_2 \sin(\beta_1 s) + a_3 \cos(\beta_2 s) + a_4 \sin(\beta_2 s). \tag{4.59}$$

Using this solution in the boundary conditions Eqs. 4.43b, c leads to the homogeneous problem $\mathbf{A}\mathbf{a} = \mathbf{0}$, where $\mathbf{a} = (a_1\ a_2\ a_3\ a_4)^T$. Vanishing $\det \mathbf{A}$ allows one to evaluate the divergence locus $\mathcal{D}$ in the (μ, ν)-plane; a significant portion of it is represented in Fig. 4.7a. For any values $(\mu_c, \nu_c) \in \mathcal{D}$ it turns out that, indeed, $(\mu_c + \nu_c)^2 > k_1$.

(b) When $\mu_c = 0$, the system is conservative. For it, two (Eulerian) critical loads ν_c are found on the locus. The lowest of them is $\nu_c = 1.325$, providing $\beta_1 = 1.481$ and $\beta_2 = 0.675$. The relevant right eigenvector, shown in Fig. 4.7b, is:

$$\phi_c(s) = a_1 \left[\cos(\beta_1 s) - \cos(\beta_2 s) - \frac{\beta_1^2 \cos\beta_1 - \beta_2^2 \cos\beta_2}{\beta_1 \sin\beta_1 - \beta_2 \sin\beta_2}\left(\frac{1}{\beta_1}\sin(\beta_1 s) - \frac{1}{\beta_2}\sin(\beta_2 s)\right)\right], \tag{4.60}$$

where $a_1 = -1.371$, in order for $\phi_c(1) = 1$. Since the boundary value problem is self-adjoint, the left eigenvector $\psi_c(s)$ coincides with $\phi_c(s)$.

As a further step, the coefficients c_{ij} are evaluated through Eqs. 4.49, which provide $c_{11} = -3.081$, $c_{12} = 11.035$, $c_{20} = -73.029$. The bifurcation diagram, as derived by Eq. 4.50, is then obtained and plotted in Fig. 4.8, where stability properties are also shown.

[22] See Exercise 2.3 for an analogous hypothesis.

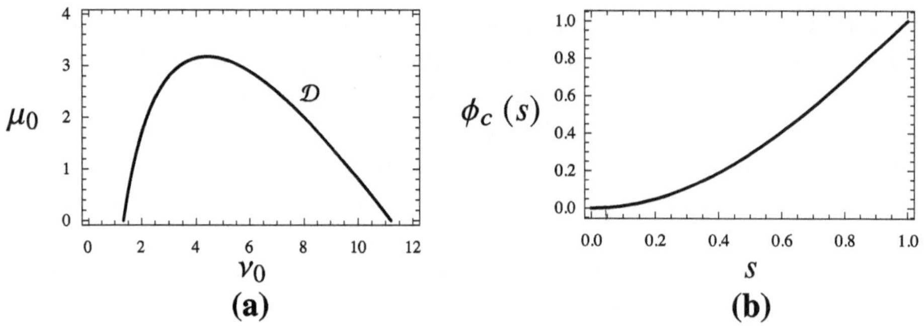

Fig. 4.7 Solution of the boundary value problem in Eq. 4.43. **a** Divergence locus $\mathcal{D}$ on the (ν, μ)-plane; **b** Right critical mode when $\mu_c = 0$, $\nu_c = 1.325$ (it coincides with the left critical mode)

Fig. 4.8 Bifurcation diagram of the beam under pure gravitational load (continuous line: stable; dashed lines: unstable)

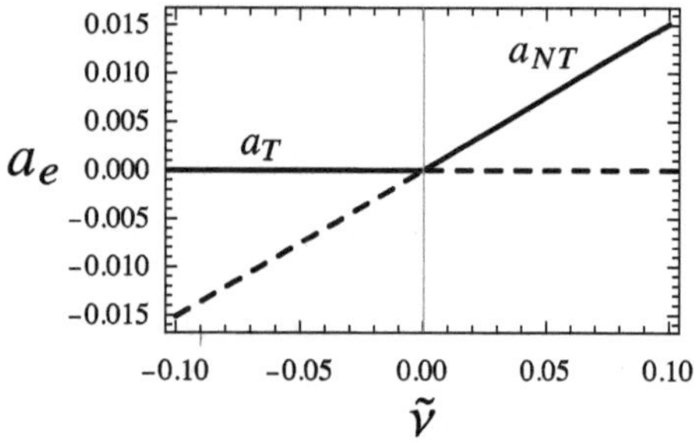

(c) When $\mu_c = 2$, the system is nonconservative. The corresponding conservative critical force is $\nu_c = 2.195$, providing $\beta_1 = 2.876$ and $\beta_2 = 0.348$. The associate critical right mode is given again by Eq. 4.59 with $a_1 = -0.076$, while the left eigenvector $\psi_c(s)$, now different from $\phi_c(s)$ (being the system non self-adjoint), is (Fig. 4.9a):

$$\begin{aligned}\psi_c(s) =& \tilde{a}_1 \Big[\cos(\beta_1 s) - \cos(\beta_2 s) \\ &- \frac{(\beta_1^2 - 4)\cos\beta_1 - (\beta_2^2 - 4)\cos\beta_2}{\beta_2(\beta_1^2 - 4)\sin\beta_1 - \beta_1(\beta_2^2 - 4)\sin\beta_2} (\beta_2 \sin(\beta_1 s) - \beta_1 \sin(\beta_2 s)) \Big],\end{aligned} \tag{4.61}$$

where $\tilde{a}_1 = -0.499$ in order for $\psi_c(1) = 1$. The coefficients c_{ij} are then evaluated through Eqs. 4.49, which provide $c_{11} = -6.353$, $c_{12} = 9.038$, $c_{20} = -68.278$. The bifurcation diagram is shown in Fig. 4.9b. In the selected range, it is close to that of the case $\mu_c = 0$. □

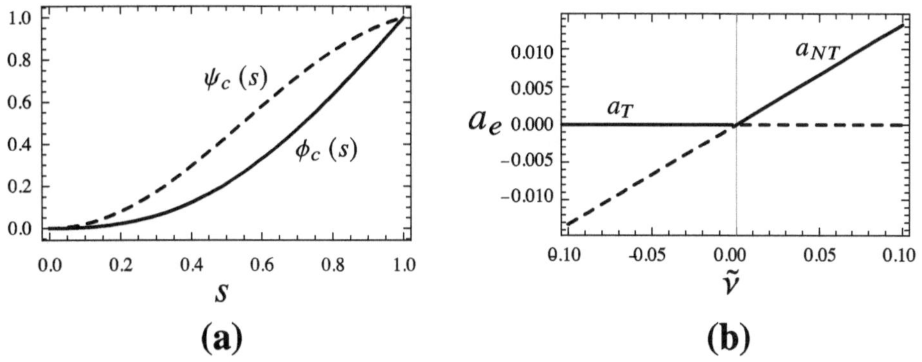

Fig. 4.9 **a** Right and left critical modes for $\mu_c = 2$, $\nu_c = 2.195$ **b** bifurcation diagram (continuous line: stable; dashed lines: unstable)

References

1. Carr, J.: Applications of Centre Manifold Theory. Springer, New York (1981)
2. Guckenheimer, J., Holmes, P.: Nonlinear Oscillations, Dynamical Systems, and Bifurcations of Vector Fields. Springer, New York (1983)
3. Luongo, A., Ferretti, M., Di Nino, S.: Stability and Bifurcation of Structures: Statical and Dynamical Systems. Springer, Cham (2023)
4. Troger, H., Steindl, A.: Nonlinear Stability and Bifurcation Theory: An Introduction for Engineers and Applied Scientists. Springer, Wien (1991)
5. Wiggins, S.: Introduction to Applied Nonlinear Dynamical Systems and Chaos. Springer, New York (1990)

Nonlinear Continuum Models 5

5.1 Introduction

To describe nonlinear mechanical phenomena, nonlinear models have to be formulated. By limiting the attention to the linear (visco-) elastic regime, nonlinearities are of geometric type. They entail nonlinear strain-displacement relationships, and call for equilibrium equations to be enforced in the current (and unknown) configuration, differently from what happens in the linear theory, where the balance of the forces is expressed in the frozen reference configuration. Alternatively, the balance law can be derived from a variational formulation, either: (a) by the Theorem of the Total Potential Energy, when the system is conservative, or (b) by the Virtual Power Principle, irrespective of the nature of the forces. These approaches only require performing a kinematic analysis and standard operations of Variational Calculus.

Here, an abstract *metamodel* is formulated, in terms of linear and nonlinear integro-differential operators, which will be useful to illustrate, in Chap. 6, how to apply perturbation algorithms to general continuous systems (methods so far applied, for introductory purposes, to the sample system only). Such a metamodel includes all (or, at least, a large class of) structural models, as string, cables, beams, membranes, plates (or even curved spatial structures, not addressed in this book), making unitary the treatment of all of them.

For any given structure, there exist in literature many nonlinear models able to describe the relevant mechanical behavior, with different level of accuracy. Very often, however, a *minimal model*, which includes few fundamental mechanical aspects, is able to capture, at least from a qualitative point of view, the main nonlinear phenomena. This philosophy of modeling is consistent with the perturbation method, which is indeed aimed to unveil in a simple way the qualitative, rather than quantitative, character of the response. In this

A. Luongo et al., *Perturbation Methods and Nonlinear Phenomena*,
Synthesis Lectures on Engineering, Science, and Technology,
https://doi.org/10.1007/978-3-031-49397-3_5

chapter, minimal models of 1D and 2D continuous structures are briefly discussed, to be used in Chap. 7 for solving specific illustrative problems.

5.2 Metamodel

All the nonlinear (visco) elastic models can be synthesized by a metamodel, i.e., an archetype, from which any models (or, at least, a class of them) can be derived as a specialization to the case at hand. Here, unprestressed systems are first considered; then, prestress is introduced.

Unprestressed systems

Reference is made to a body (or a collection of bodies), made of viscoelastic material, occupying, in the reference configuration, a (1D, 2D or 3D) domain Ω,[1] of boundary Γ. Its equations of motion take the form:

$$\mathcal{M}\ddot{\mathbf{u}} + \mathcal{C}\dot{\mathbf{u}} + \mathcal{K}\mathbf{u} + \boldsymbol{n}(\mathbf{u}) = \mathbf{b}, \qquad \text{in } \Omega, \tag{5.1a}$$

$$\mathcal{D}\dot{\mathbf{u}} + \mathcal{B}\mathbf{u} = \mathbf{f}, \qquad \text{on } \Gamma. \tag{5.1b}$$

Here: $\mathbf{u} = \mathbf{u}(\mathbf{s}, t)$ is the displacement field, measured from the reference configuration, depending on the position $\mathbf{s}$ and time t, and a dot denotes time-differentiation; $\mathcal{M}, \mathcal{C}, \mathcal{K}$ are the mass, damping and stiffness linear differential (or integro-differential) operators, respectively, acting in the domain; $\mathcal{D}, \mathcal{B}$ are differential operators acting at the boundary, collecting both geometric and mechanical conditions; $\boldsymbol{n}(\mathbf{u})$, with $\boldsymbol{n}(\mathbf{0}) = \mathbf{0}$, is a *vectors of nonlinearities* in the domain,[2] which depend on $\mathbf{u}$ and their spatial- and/or time-derivatives, or even on their integrals; $\mathbf{b} = \mathbf{b}(\mathbf{s}, t)$ and $\mathbf{f} = \mathbf{f}(\mathbf{s}, t)$ are assigned body and surface forces, respectively.

If the exact, nonlinear equations of motion are known, in order to put them in the form of Eq. 5.1, a series expansion of $\mathbf{u}$ must be carried out around the trivial configuration $\mathbf{u} = \mathbf{0}$. This operation allows one to separate the linear part of the equation from the nonlinear part. It, moreover, leads to express the nonlinearities as a sum of polynomial contributions, of the type:

$$\boldsymbol{n}(\mathbf{u}) = \boldsymbol{n}_2(\mathbf{u}, \mathbf{u}) + \boldsymbol{n}_3(\mathbf{u}, \mathbf{u}, \mathbf{u}) + \cdots, \tag{5.2}$$

where $\boldsymbol{n}_k,\ k = 2, 3, \ldots$ are quadratic, cubic, ..., homogeneous forms of their arguments.

Prestressed systems

If the external forces enter the left hand member of the equations, being multiplied by the configuration variables, it is customary to refer to them as *presolicitation forces*, assumed to be proportional to a load multiplier μ.[3] These forces also contribute to the stiffness and

[1] The symbol Ω, denoting the domain, should not be confused with $\mathit{\Omega}$, extensively used throughout the book, to denote external or parametric frequencies.

[2] A more general class of systems should include nonlinearities at the boundary, considered in this book only for the model in Sect. 5.5.2.

[3] Here, no distinction is made between multipliers μ, ν_s, ν_d used in Chaps. 1–4.

damping operators, as well as to the nonlinearity vector. Accordingly, $\mathcal{K} = \mathcal{K}_e + \mu\mathcal{K}_g$, $\mathcal{C} = \mathcal{C}_s + \mu\mathcal{C}_g$, where $\mathcal{K}_e, \mathcal{C}_s$ are the *elastic stiffness* operator and the *structural damping* operator, respectively, evaluated at $\mathbf{u} = \mathbf{0}$; moreover, $\mathcal{K}_g, \mathcal{C}_g$ are their *geometric* counterparts, which account for the linear part of the external presolicitation forces. The same occurs at the boundary. Concerning nonlinearities, it is assumed that $\boldsymbol{n}(\mathbf{0}; \mu) = \mathbf{0}$ for any μ.[4] Examples are offered by the compressed Euler beam, in which the load μ enters the stiffness, or an aeroelastic system, in which the wind flow μ affects the damping matrix. In these cases, the Eqs. 5.1 assume the form:

$$\mathcal{M}\ddot{\mathbf{u}} + \left(\mathcal{C}_s + \mu\mathcal{C}_g\right)\dot{\mathbf{u}} + \left(\mathcal{K}_e + \mu\mathcal{K}_g\right)\mathbf{u} + \boldsymbol{n}(\mathbf{u}; \mu) = \mathbf{b}, \qquad \text{in } \Omega, \tag{5.3a}$$

$$\left(\mathcal{D}_s + \mu\mathcal{D}_g\right)\dot{\mathbf{u}} + \left(\mathcal{B}_s + \mu\mathcal{B}_g\right)\mathbf{u} = \mathbf{f}, \qquad \text{on } \Gamma. \tag{5.3b}$$

In what follows, a gallery of examples will be browsed, showing as models of string, cables, beams, membranes and plates fall in the class of systems discussed above. In all the examples, however, the exact equations of motion will be not derived, as they can be found in specialized texts, but approximate *minimal models*, still able to capture the main qualitative behavior, will be described, amenable to be handled by analytical methods, as in the spirit of this book.[5] Basic concepts of structural analysis are assumed to be known to the reader.

5.3 Taut String

A planar, undamped, elastic string, of mass per unit length m, taut by a presolicitation axial force N_0 (positive when of traction), fixed at the ends $s = 0, \ell$, and loaded by transverse forces $p_y = p_y(s, t)$, is considered. The equation expressing the funicular equilibrium reads[6] [5, 6]:

$$m\ddot{v} - \left[\left(N_0 + \tilde{N}\right)v'\right]' = p_y, \tag{5.4}$$

where $v = v(s, t)$ is the transverse displacement and $\tilde{N} = \tilde{N}(s, t)$ the dynamic tension. The elastic law requires that $\tilde{N} = EA\varepsilon$, where EA is the axial stiffness of the string and $\varepsilon = \varepsilon(s, t)$ the unit extension. Under the hypothesis $u \ll v$, kinematics establishes that $\varepsilon = u' + \frac{1}{2}v'^2$,[7] in which $u = u(s, t)$ is the longitudinal displacement. The problem is completed by the longitudinal equation of motion, $\tilde{N}' - m\ddot{u} = 0$. If, however, one is interested in

[4] If this is not the case, a change of variable allows recasting the equations in this form [4].

[5] Minimal models are not always self-consistent, in the sense that some nonlinear terms are missed, of the same power of $\mathbf{u}$ of those retained, since believed smaller.

[6] The funicular effect is due to the rotation v' of the presolicited material element, which induces the birth of a geometric transverse force $\left(N_0 + \tilde{N}\right)v'$, whose derivative is a carrying load.

[7] It can be derived by the Green-Lagrange strain measure, or expanding the exact expression of the unit extension $\varepsilon = \sqrt{(1 + u')^2 + v'^2} - 1$.

prevalently transverse motions, the longitudinal inertia forces can be neglected, since the (first) transverse frequencies are much smaller than the longitudinal ones, this entailing a small dynamic coupling. This simplification allows ones to assume $\tilde{N} = \tilde{N}(t)$, i.e., the dynamic tension is taken, at each time, *constant along the string*. From the elasto-geometric equation and longitudinal boundary conditions, i.e.:

$$u' = \frac{\tilde{N}(t)}{EA} - \frac{1}{2}v'^2, \tag{5.5a}$$

$$u(0) = 0, \tag{5.5b}$$

$$u(\ell) = 0, \tag{5.5c}$$

u can be expressed as dependent on v; this operation is known as *static condensation* of the longitudinal displacements. By integrating Eq. 5.5a in $(0, \ell)$, a compatibility condition follows, namely:

$$\tilde{N}(t) = \frac{1}{2}\frac{EA}{\ell}\int_0^\ell v'^2 \mathrm{d}s. \tag{5.6}$$

When this is substituted in Eq. 5.4 and the transverse boundary conditions are appended, the final problem reads:

$$m\ddot{v} - \left(N_0 + \frac{1}{2}\frac{EA}{\ell}\int_0^\ell v'^2 \mathrm{d}s\right) v'' = p_y, \tag{5.7a}$$

$$v(0, t) = 0, \tag{5.7b}$$

$$v(\ell, t) = 0. \tag{5.7c}$$

It is noticed that, with the symbols in Eq. 5.3, $\mathcal{M} := m$, and the stiffness is only of geometric type, i.e., $\mathcal{K}_e := 0$, $\mathcal{K}_g := -N_0\frac{\partial^2}{\partial s^2}$. Moreover, the nonlinearities are cubic (i.e., odd, due to the symmetry of the system), of integro-differential type: $n_2(v, v) := 0$, $n_3(v, v, v) := -\frac{1}{2}\frac{EA}{\ell}v'' \int_0^\ell v'^2 \mathrm{d}s$. An application relevant to the statics of the string is presented in Problem 7.1.

5.4 Suspended Cable

An undamped planar suspended cable, hanging from two fixed supports placed at the same level, is considered. Due to the own weigh, of intensity mg per unit length, with m the mass per unit length and g the gravity acceleration, the cable lies on an arc of catenary. However, if its sag d is small with respect the span ℓ (or, equivalently, if its length is only slightly larger than the distance between the supports), such a curve can be approximated by a parabola [1, 2]. Consistently, the slope of the curved profile is assumed small, the curvature k_0 and the static tension N_0 are taken nearly constant along the cable, related by $k_0 := \frac{mg}{N_0} = \frac{8d}{\ell^2}$.[8] Such an assumption is known as the *approximation of the parabolic cable* or *of the shallow cable.*

The equation expressing the funicular equilibrium reads [1–3, 5]:

$$\left(N_0 + \tilde{N}\right) v'' + \tilde{N} k_0 + p_y = m\ddot{v}, \tag{5.8}$$

which generalizes, with the same meaning of symbols, the equation of the taut string, Eq. 5.4.[9]

The elastic law requires that $\tilde{N} = EA\varepsilon$, where EA is the axial stiffness of the string and, from kinematics, $\varepsilon = u' - k_0 v + \frac{1}{2} v'^2$ is the unit extension, which accounts for the existing curvature of the cable.[10]

To complete the problem, the longitudinal equation of motion must be added. By exploiting the hypothesis of shallow cable, it can still be taken as for the string, i.e., $\tilde{N}' - m\ddot{u} = 0$. By neglecting, once again, the longitudinal inertia forces, $\tilde{N} = \tilde{N}(t)$ follows, entailing a change of tension, at time t, constant along the cable. From the elasto-geometric equation $u' = \frac{\tilde{N}(t)}{EA} + k_0 v - \frac{1}{2} v'^2$, by integrating and enforcing $u(0) = u(\ell) = 0$, the dynamic tension is found:

$$\tilde{N}(t) = \frac{EA}{\ell} \left(-k_0 \int_0^\ell v \mathrm{d}s + \frac{1}{2} \int_0^\ell v'^2 \mathrm{d}s \right). \tag{5.9}$$

The transverse equation of motion, Eq. 5.8, can therefore be expressed in terms of transverse displacement only; by appending the boundary conditions, it reads:

$$\begin{aligned} m\ddot{v} - N_0 v'' &+ \frac{EA}{\ell} k_0^2 \int_0^\ell v \mathrm{d}s - \frac{EA}{2\ell} k_0 \int_0^\ell v'^2 \mathrm{d}s \\ &+ \frac{EA}{\ell} k_0 v'' \int_0^\ell v \mathrm{d}s - \frac{EA}{2\ell} v'' \int_0^\ell v'^2 \mathrm{d}s = p_y, \end{aligned} \tag{5.10a}$$

$$v(0, t) = 0, \tag{5.10b}$$

$$v(\ell, t) = 0. \tag{5.10c}$$

[8] The relation $N_0 d = \frac{1}{8} m g \ell^2$ expresses the global rotational equilibrium of half a cable.

[9] The Eq. 5.8 can be drawn by starting from Eq. 5.4, but by adding the preexisting curvature k_0 to the increment of curvature v'', as well as the pre-existing external load $-mg$ to the incremental forces p_y, thus obtaining:

$$m\ddot{v} - \left(N_0 + \tilde{N}\right)\left(k_0 + v''\right) = p_y - mg.$$

Since, for static equilibrium, is $N_0 k_0 - mg = 0$, Eq. 5.8 follows.

[10] Indeed, a circumference, whose initial radius k_0^{-1} is reduced by v, undergoes an unitary extension $\varepsilon = -k_0 v$.

It is seen that, in addition to those relevant to the taut string, new elastic terms appear in integral form: (i) a linear elastic contribution $\mathcal{K}_e v := \frac{EA}{\ell} k_0^2 \int_0^\ell v \mathrm{d}s$, (ii) a quadratic nonlinearity $n_2(v, v) := -\frac{1}{2}\frac{EA}{\ell} k_0 \int_0^\ell v'^2 \mathrm{d}s + \frac{EA}{\ell} k_0 \int_0^\ell v \mathrm{d}s\, v''$, which expresses the lack of symmetry induced by the curvature (manifesting in the nonlinear range, only). An application regarding the free dynamics of the cable is illustrated in Problem 7.4.

Remark 5.1 It should be noticed in Eq. 5.10, that if v is antisymmetric with respect the midspan, the linear elastic term disappears, due to the fact that, for the approximations introduced, the linear part of the total elongation vanishes.

5.5 Beams

To analyze beams in the regime of large transverse displacements, two nonlinear geometric effects have to be accounted for: (a) the bending curvature is a nonlinear function of the displacements; (b) the beam undergoes axial elongations (as already observed for the taut string). Aimed to formulate simple models, it is customary to distinguish two different categories of beams, according to the nature of their constraints at the ends, namely: (a) beams with *axial constrains at both ends* (e.g., clamped-clamped, hinged-hinged), and (b) beams with *one axial constraint* (e.g., hinged-rolled, clamped-free, clamped-sliding).

When the constraints do not slide, the beam suffers large elongations, whose elastic energy contribution is much larger of that relevant to the nonlinear curvature. Therefore, one can, with reasonable approximation, to consider *linear curvatures and nonlinear elongations*, for which reasoning similar to those developed for the taut string hold. When, in contrast, the constraints allow the sliding, the elongations of the beam are negligible (suggesting the beam can be considered inextensible), so that the main source of nonlinearity is due to the curvature. Beams of the first category deviate much more from linearity than beams of the second category. As a matter of fact, the nonlinear mechanical behavior of the former is appreciated when the amplitude of the response is of the order of the thickness of the beam; of the latter, when the response is, e.g., of the order of 1/10 of the length, i.e., of the order of 10 times the thickness, for beams with length-to-thickness ratio of the order of 100.

The two classes of beams are separately studied in the next sub-sections.

5.5.1 Axially Restrained Beam

Axially restrained planar beams are here modeled as Euler-Bernoulli beams, for which the curvature is assumed to be small, expressed as the second derivative of the transverse displacement, $\kappa = v''$ (as in the linear theory). In contrast, the extension is assumed to be large, expressed by $\varepsilon = u' + \frac{1}{2}v'^2$. The transverse equilibrium equation is easily obtained by adding up two bearing capacities[11]: (a) the *flexural bearing capacity*, EIv'''', where EI is its flexural stiffness, and (b) the *funicular bearing capacity* $-Nv''$, where $N = N_0 + \tilde{N}$ is the axial force, sum of a (possible) prestress $N_0 = \text{const}$ and of a (constant along the beam) dynamic component $\tilde{N} = \tilde{N}(t)$, given, in the approximations already discussed, by Eq. 5.6. By introducing external and internal damping forces, $-c_e\dot{v}$ and $\eta I\dot{v}''''$, respectively, with c_e and η viscosity coefficients, the equation of motion reads[12]:

$$m\ddot{v} + c_e\dot{v} + EIv'''' + \eta I\dot{v}'''' - \left(N_0 + \frac{EA}{2\ell}\int_0^\ell v'^2 \mathrm{d}s\right) v'' = p_y, \tag{5.11}$$

to which proper boundary conditions for transverse displacement must be appended. If, e.g., the beam is hinged-hinged, they are:

$$v(0,t) = 0, \tag{5.12a}$$

$$v(\ell,t) = 0, \tag{5.12b}$$

$$EIv''(0,t) + \eta I\dot{v}''(0,t) = 0, \tag{5.12c}$$

$$EIv''(\ell,t) + \eta I\dot{v}''(\ell,t) = 0, \tag{5.12d}$$

in which the last two equations express the vanishing of the viscoelastic bending moment $M = EIv'' + \eta I\dot{v}''$ at the ends.

With the formalism of Eq. 5.3, new linear operators appear with respect to the string, namely: $\mathcal{C}_s := c_e + \eta I\frac{\partial^4}{\partial s^4}$, $\mathcal{K}_e := EI\frac{\partial^4}{\partial s^4}$ in the field, and $\mathcal{D}_s := \eta I\frac{\partial^2}{\partial s^2}$, $\mathcal{B}_e := EI\frac{\partial^2}{\partial s^2}$ at the boundary (the latter two concerning the mechanical conditions only). Applications concerning forced and free dynamics of the axially restrained beam are discussed in the Problems 7.7, 7.8.

5.5.2 Axially Unrestrained Beam

When the (Euler-Bernoulli) beam is axially unrestrained, as already commented, it is convenient to resort to the inextensible model, for which $\varepsilon = u' + \frac{1}{2}v'^2 = 0$. This geometric internal constraint, allows us to express the longitudinal displacement u as a function of the

---

[11] The beam, therefore, behaves an in-parallel system, whose components are: (i) a purely flexural beam, and (ii) a string.

[12] The external damping forces express the action exerted by the air on the beam; they are proportional to the local velocity via the damping coefficient c_e. The internal damping is modeled according to the Kelvin-Voigt rheological model (made of an in-parallel spring E and a dashpot η), for which the stress-strain relationships reads $\sigma = E\varepsilon + \eta\dot{\varepsilon}$. This leads to the so called *Similarity Principle*, according to which $E \to E + \eta\frac{\partial}{\partial t}$, which permits to relate a purely elastic and a viscoelastic model.

transverse displacement v. By assuming that $u(0,t)=0$, while $u(\ell,t)\neq 0$, by integrating the constraint, it follows:

$$u=-\frac{1}{2}\int_0^s v'^2 \mathrm{d}s. \tag{5.13}$$

Therefore, once again, the condensation is successful, although based on a kinematic, rather than an equilibrium condition. Moreover, being the curvature κ large, a nonlinear kinematic relationship is needed to express it. Since, from elementary geometric consideration, it is $v'=\sin\theta$,[13] with θ the rotation of the beam axis, $\kappa:=\theta'$ assumes the exact expression $\kappa=\frac{v''}{\sqrt{1-v'^2}}$. When it is expanded in series:

$$\kappa=v''\left(1+\frac{1}{2}v'^2\right). \tag{5.14}$$

The inextensibility constraint entails two important consequences on the internal forces: (a) the dynamic part of the axial force, $\tilde{N}$, assumes a reactive character, so that it cannot derived from the constitutive law, but rather from equilibrium; (b) the longitudinal inertia force, $-m\ddot{u}$, is no more negligible, implying that $\tilde{N}$ is variable along s, differently from the axially restrained beam.

The derivation of the equation of motion is non-trivial, as in the previous cases, but requires some care. The beam is considered to be loaded by distributed transverse forces $p_y(s,t)$ and a concentrated compressive force applied at the right end, either gravitational, P, or follower, F. The field balance equations, when projected onto the reference basis, read:

$$H'=m\ddot{u}, \tag{5.15a}$$

$$V'=m\ddot{v}+c_e\dot{v}-p_y, \tag{5.15b}$$

$$M'+V\cos\theta-H\sin\theta=0, \tag{5.15c}$$

where H, V are the longitudinal (horizontal) and transverse (vertical) components of the internal reactive force, and M is the bending moment. The first of them can be integrated to furnish:

$$H=m\int_\ell^s \ddot{u}\mathrm{d}s+H_\ell, \tag{5.16}$$

where $H_\ell:=H(\ell,t)$. It is, either, $H_\ell=-P$ or $H_\ell=-F\cos(\theta(\ell,t))$, for the two different forces. Since, from Eq. 5.15c, it is $V=H\tan\theta-\frac{M'}{\cos\theta}$, Eq. 5.15b reads:

$$\left(H\tan\theta-\frac{M'}{\cos\theta}\right)'=m\ddot{v}+c_e\dot{v}-p_y. \tag{5.17}$$

---

[13] Indeed, an inextensible segment ds which rotates of θ has transverse projection $\mathrm{d}v=\mathrm{d}s\sin\theta$.

For small displacements it is: $\cos\theta = \sqrt{1-\sin^2\theta} \simeq 1 - \frac{1}{2}v'^2$ and $\tan\theta = \frac{v'}{\sqrt{1-v'^2}} \simeq v'\left(1+\frac{1}{2}v'^2\right)$. By using the constitutive law $M = EI\kappa + \eta I\dot{\kappa}$ for viscoelastic material, Eq. 5.14 for the curvature and Eq. 5.13 to express the longitudinal inertia force, Eq. 5.17 is finally recast as:

$$m\ddot{v} + c_e\dot{v} + EI\left\{v'''' + \left[v'\left(v'v''\right)'\right]'\right\} + \eta I\dot{v}'''' - H_\ell v''\left(1+\frac{3}{2}v'^2\right) + \frac{1}{2}m\left[v'\int_\ell^s\left(\int_0^s\left(v'^2\right)^{\cdot\cdot}\mathrm{d}s\right)\mathrm{d}s\right]' = p_y, \tag{5.18}$$

in which nonlinear viscous terms have been neglected.[14] If the end load is gravitational, the geometric term is $Pv''\left(1+\frac{3}{2}v'^2\right)$; if it is follower, since $H_\ell \simeq -F\left(1-\frac{1}{2}v'^2(\ell,t)\right)$, the geometric term reads $Fv''\left(1+\frac{3}{2}v'^2(s,t)-\frac{1}{2}v'^2(\ell,t)\right)$ plus higher order terms.

Proper boundary conditions for the transverse displacement must be appended. The geometric conditions are trivial, namely $v=0$ and/or $\theta=0$, this latter entailing $v'=0$, as for the linear case. The mechanical conditions, instead, are nonlinear, since they call for expressing the internal forces in terms of displacements, consistently with the approximations introduced. If, e.g., the right end of the beam is free to rotate, $M(\ell,t) = EIv''(\ell,t)\left(1+\frac{1}{2}v'^2(\ell,t)\right) + \eta I\dot{v}''(\ell,t) = 0$ must be enforced, in which the nonlinear viscous terms have been neglected. If, in addition, a follower compression force F is there applied, the condition $V(\ell,t) = -F\sin(\theta(\ell,t))$ must hold. Since, from Eq. 5.15c, $V(\ell,t) = H_\ell\tan\theta(\ell,t) - \frac{M'}{\cos\theta(\ell,t)}$ and $H_\ell = -F\cos(\theta(\ell,t))$, the mechanical boundary condition simplifies into $-\frac{M'(\ell,t)}{\cos(\theta(\ell,t))} = 0$, i.e., $M'(\ell,t)=0$.

It should be noticed in this model, that the nonlinearities are elastic (due to the curvature) as well as inertial, due to the sliding movement; accordingly, $n_3(v,v,v) := EI\left[v'\left(v'v''\right)'\right]' + \frac{1}{2}m\left[v'\int_\ell^s\left(\int_0^s(v'^2)^{\cdot\cdot}\,\mathrm{d}s\right)\mathrm{d}s\right]'$. Moreover, geometric nonlinearities are present, namely, $\frac{3}{2}Pv'^2v''$ in the gravitational case, or $\frac{1}{2}Fv''\left(3v'^2(s,t) - v'(\ell,t)^2\right)$ in the follower case. Nonlinearities appear also at the boundary, as a generalization of the meta-model in Eq. 5.3.

Dynamic problems regarding axially unrestrained beams are discussed in the Problems 7.5, 7.9; a dynamic bifurcation problem is illustrated in the Problem 7.10.

5.6 Taut Membrane

As an example of two dimensional continuous system, an undamped elastic membrane is considered, which, in the reference configuration, is presolicited by a hydrostatic plane stress state, of intensity N_0 (force per unit length). It occupies a domain Ω of the (x,y) plane, and

[14] Moreover, the following identity has been exploited: $v''''v'^2 + 4v'v''v''' + v''^3 = \left[v'\left(v'v''\right)'\right]'$.

it is fixed along the whole boundary Γ.[15] The membrane undergoes out-of-plane (along z) displacements $w(x, y, t)$ and in-plane displacements $u(x, y, t)$, $v(x, y, t)$, in the x and y directions, respectively.

Equilibrium
The balance of the out-of-pane forces states that:

$$m\ddot{w} - [(N_0 + \tilde{N}_x)w_{,x}]_{,x} - [(N_0 + \tilde{N}_y)w_{,y}]_{,y} - [\tilde{N}_{xy}w_{,x}]_{,y} - [\tilde{N}_{xy}w_{,y}]_{,x} = p_z, \tag{5.19}$$

in which m is the mass per unit of area, $p_z = p_z(x, y, t)$ is the load in the z direction and a comma denotes differentiation with respect the following variable; moreover, $\tilde{N}_{\alpha\beta} = \tilde{N}_{\alpha\beta}(x, y, t)$ $(\alpha, \beta = x, y)$, are in-plane dynamic increments of the normal $(\tilde{N}_x, \tilde{N}_y)$ and shear $(\tilde{N}_{xy})$ stresses. Equation 5.19 generalizes to two dimensions the Eq. 5.4, holding in the one dimensional case, being based on the same mechanism of the funicular equilibrium.[16]

If the interest is focused on prevalently transverse vibrations, the in-plane inertial forces $m\ddot{u}, m\ddot{v}$ can be neglected, as done for the string. Accordingly, the dynamic stresses must satisfy static equilibrium conditions, namely:

$$\tilde{N}_{x,x} + \tilde{N}_{xy,y} = 0, \qquad \text{in } \Omega, \tag{5.20a}$$
$$\tilde{N}_{xy,x} + \tilde{N}_{y,y} = 0, \tag{5.20b}$$

in which it has been assumed, that no in-plane external loads act on the structure. Differently from the case of the string, however, and due to the increased dimensions of the problem (which is now internally hyperstatic), the membrane equilibrium Eqs. 5.20 *cannot be integrated to allow condensation* of the in-plane displacements. Moreover, $N_{\alpha\beta}$ are not more constant in the domain, as for the string. In spite of this last consideration, by accounting for the in-plane equilibrium Eqs. 5.20, the out-of-plane equilibrium Eqs. 5.19 simplify into[17]:

$$m\ddot{w} - \left(N_0 + \tilde{N}_x\right) w_{,xx} - \left(N_0 + \tilde{N}_y\right) w_{,yy} - 2\tilde{N}_{xy}w_{,xy} = p_z. \tag{5.21}$$

---

[15] The condition of zero displacements at the boundary will be later relaxed, concerning their component tangent to Γ.

[16] The first two geometric contributions to the equilibrium are the *funicular carrying loads* of two mutually orthogonal strings. The third term, $\tilde{N}_{xy}w_{,x}$ represents the component along z of the shear stress $\tilde{N}_{xy}$, caused by the rotation of the element around the y axis, whose derivative with respect to y is a *torsional geometric carrying load*. The same occurs for the fourth term.

[17] That is, as $N_{\alpha\beta}$ were (formally) constant.

Displacement method

To eliminate the incremental stresses from this equation, the displacement method is followed here.[18] Stresses are expressed in terms of strains, according to the hypothesis of plane state of stress, i.e.:

$$\begin{pmatrix} \tilde{N}_x \\ \tilde{N}_y \\ \tilde{N}_{xy} \end{pmatrix} = C \begin{pmatrix} 1 & \nu & 0 \\ \nu & 1 & 0 \\ 0 & 0 & \frac{1-\nu}{2} \end{pmatrix} \begin{pmatrix} \varepsilon_x \\ \varepsilon_y \\ \gamma_{xy} \end{pmatrix}, \tag{5.22}$$

where $C := \frac{Eh}{1-\nu^2}$ is the membrane stiffness, in which h is the thickness, E the Young modulus and ν the Poisson ratio. The strains, in turn, are expressed in terms of displacements, by using truncated expressions of the Green-Lagrange strains, according to the hypotheses $u, v \ll w$,[19]:

$$\varepsilon_x = u_{,x} + \frac{1}{2} w_{,x}^2, \tag{5.23a}$$

$$\varepsilon_y = v_{,y} + \frac{1}{2} w_{,y}^2, \tag{5.23b}$$

$$\gamma_{xy} = u_{,y} + v_{,x} + w_{,x} w_{,y}. \tag{5.23c}$$

By using Eqs. 5.22 and 5.23, the out-of-plane equilibrium Eqs. 5.21 become:

$$\begin{aligned} m\ddot{w} - N_0(w_{,xx} + w_{,yy}) - C\Big[& u_{,x} w_{,xx} + v_{,y} w_{,yy} + \nu u_{,x} w_{,yy} + \nu v_{,y} w_{,xx} \\ & + (1-\nu) u_{,y} w_{,xy} + (1-\nu) v_{,x} w_{,xy} \Big] - C\Big[\frac{1}{2} w_{,x}^2 w_{,xx} + \frac{1}{2} w_{,y}^2 w_{,yy} \\ & + \frac{1}{2} \nu w_{,y}^2 w_{,xx} + \frac{1}{2} \nu w_{,x}^2 w_{,yy} + (1-\nu) w_{,x} w_{,y} w_{,xy} \Big] = p_z. \end{aligned} \tag{5.24}$$

Similarly, the in-plane equilibrium Eqs. 5.20 transform into:

$$u_{,xx} + \check{\nu} u_{,yy} + \hat{\nu} v_{,xy} + w_{,x} w_{,xx} + \check{\nu} w_{,x} w_{,yy} + \hat{\nu} w_{,y} w_{,xy} = 0, \tag{5.25a}$$

$$\hat{\nu} u_{,xy} + v_{,yy} + \check{\nu} v_{,xx} + \check{\nu} w_{,y} w_{,xx} + w_{,y} w_{,yy} + \hat{\nu} w_{,x} w_{,xy} = 0, \tag{5.25b}$$

where $\hat{\nu} := \frac{1+\nu}{2}$ and $\check{\nu} := \frac{1-\nu}{2}$ has been introduced, for notation convenience. Equations 5.24, 5.25 constitute a differential system in the $\mathbf{u} := (u, v, w)$ unknowns, to be integrated with the boundary conditions on Γ. A simple case occurs when only normal displacements are restrained at the boundary, requiring the stresses tangent to Γ vanish.

---

[18] An alternative approach is offered by the use of the Airy stress function, in the context of the force methods [7].

[19] These approximate strains are known as the 'Von Karman strains'.

It should be noticed, that both quadratic and cubic nonlinearities are present in these equations, so that $\boldsymbol{n}_2(\mathbf{u}, \mathbf{u}) \neq \mathbf{0}$, $\boldsymbol{n}_3(\mathbf{u}, \mathbf{u}, \mathbf{u}) \neq \mathbf{0}$ in Eqs. 5.1, 5.2. An example relevant to the statics of the membrane is worked out in the Problem 7.2.

5.7 In-Plane Restrained Plate

A viscoelastic externally damped plate is considered, occupying a domain Ω of the (x, y) plane of boundary Γ, presolicited by a homogeneous in-plane state of stress $\left(N_x^0, N_y^0, N_{xy}^0\right)$. The in-plane $u(x, y, t)$, $v(x, y, t)$ displacements are suitably restrained along Γ, while out-of-plane displacements $w(x, y, t)$ are constrained in arbitrary manner. When the plate undergoes prevalent out-of-plane displacements, large in-plane elongations and shear strains arise as nonlinear effect, whose elastic energy contribution prevails on that related to the nonlinear curvature, as already observed for restrained beams. Therefore, it is reasonable to model the flexural and torsional curvatures of the plate as linear, while the membrane strains are taken nonlinear. This consideration leads to write the balance of out-of-plane forces, by simply adding the in-parallel linear stiffness of the plate to the geometrical stiffness of the taut membrane Eq. 5.19 (with updated state of prestress). By accounting for the external damping and the visco-elastic nature of the material, this equation reads:

$$
\begin{aligned}
m\ddot{w} + c_e\dot{w} + D\nabla^4 w + D_v\nabla^4\dot{w} - [(N_x^0 + \tilde{N}_x)w_{,x}]_{,x} - [(N_y^0 + \tilde{N}_y)w_{,y}]_{,y} \\
-[(N_{xy}^0 + \tilde{N}_{xy})w_{,x}]_{,y} - [(N_{xy}^0 + \tilde{N}_{xy})w_{,y}]_{,x} = p_z,
\end{aligned}
\tag{5.26}
$$

where c_e is the external damping coefficient, $D := \frac{Eh^3}{12(1-\nu^2)}$ is the flexural elastic stiffness of the plate, $D_v := \frac{\eta h^3}{12(1-\nu^2)}$ the flexural viscous impedance (with η the internal damping of the material).

The balance of the in-plane forces should incorporate the viscous and damping forces in Eqs. 5.20. However, by neglecting any dynamic effect, as already done for the inertial forces, those equations can still be considered valid for the viscoelastic plate. This entail that Eq. 5.26 can be simplified as follows:

$$
\begin{aligned}
m\ddot{w} + c_e\dot{w} + D\nabla^4 w + D_v\nabla^4\dot{w} - (N_x^0 + \tilde{N}_x)w_{,xx} - (N_y^0 + \tilde{N}_y)w_{,yy} \\
-2(N_{xy}^0 + \tilde{N}_{xy})w_{,xy} = p_z.
\end{aligned}
\tag{5.27}
$$

When this latter is expressed in terms of displacements, the equation reads:

$$
\begin{aligned}
&m\ddot{w}+c_e\dot{w}+D\nabla^4 w+D_v\nabla^4\dot{w}-N_x^0 w_{,xx}-N_y^0 w_{,yy}-2N_{xy}^0 w_{,xy}\\
&-C\Big[u_{,x}\,w_{,xx}+v_{,y}\,w_{,yy}+\nu\,u_{,x}\,w_{,yy}+\nu\,v_{,y}\,w_{,xx}+(1-\nu)u_{,y}\,w_{,xy}\\
&+(1-\nu)v_{,x}\,w_{,xy}\Big]-C\left[\frac{1}{2}w_{,x}^2\,w_{,xx}+\frac{1}{2}w_{,y}^2\,w_{,yy}+\frac{1}{2}\nu\,w_{,y}^2\,w_{,xx}\right.\\
&\left.+\frac{1}{2}\nu\,w_{,x}^2\,w_{,yy}+(1-\nu)\,w_{,x}\,w_{,y}\,w_{,xy}\right]=p_z,
\end{aligned}
\tag{5.28}
$$

to which the in-plane equilibrium conditions Eqs. 5.25 must be appended, together with the boundary conditions.

As for the membrane, both quadratic and cubic nonlinearities appear in the plate model. Buckling and postbuckling of a plate are studied in the Problem 7.3; forced dynamics in the Problem 7.6.

References

1. Irvine, H.M., Caughey, T.K.: The linear theory of free vibrations of a suspended cable. Proc. R. Soc. Lond. A **341**(1626), 299–315 (1974)
2. Irvine, M.: Cable Structures. Dover, New York (1992)
3. Lacarbonara, W.: Nonlinear Structural Mechanics: Theory, Dynamical Phenomena and Modeling. Springer Science & Business Media, New York (2013)
4. Luongo, A., Ferretti, M., Di Nino, S.: Stability and Bifurcation of Structures: Statical and Dynamical Systems. Springer, Cham (2023)
5. Luongo, A., Zulli, D.: Mathematical Models of Beams and Cables. Wiley, New York (2013)
6. Nayfeh, A.H., Mook, D.T.: Nonlinear Oscillations. Wiley, New York (1995)
7. Timoshenko, S.P., Woinowsky-Krieger, S.: Theory of Plates and Shells. McGraw-Hill, New York (1959)

Perturbation Methods for a Continuum Metamodel

6

6.1 Introduction

The sample model (i.e., the beam on elastic soil) was useful to illustrate all the algorithmic aspects relevant to the perturbation methods. The calculations were simplified by the circumstance that all the nonlinearities are, in that case, of algebraic type. The extension to more general (and realistic) systems is, however, straightforward. Indeed, although the nonlinearities are now of differential (or integro-differential) nature, they do not change the algorithm itself, since, in the perturbation scheme, they must be evaluated as 'know terms', depending on lower-order solutions. In this way, the reader who has learned how to solve problems concerning the sample model, is able to solve problems relevant to *any* continuous model, although with much heavier computations. To stress this property, use is made in this chapter of the metamodel introduced in Chap. 5, for which all the perturbation algorithms discussed in Chaps. 2–4, are revisited.

6.2 Nonlinear Elastostatics

The nonlinear elastostatic problem, already discussed in the Sect. 2.2 for the sample system, is now tackled for a general system. The straightforward perturbation method is used [12].

An elastic structure, isostatically or hyperstatically constrained at the ground, subject to gravitational loads, is considered in the static regime. The relevant equilibrium equations are derived from Eq. 5.1, by neglecting any dynamic effect and taking $\mathbf{u} = \mathbf{u}(\mathbf{s})$, i.e.:

$$\mathcal{K}\mathbf{u} = -\boldsymbol{n}_2(\mathbf{u}, \mathbf{u}) - \boldsymbol{n}_3(\mathbf{u}, \mathbf{u}, \mathbf{u}) + \cdots + \alpha\mathbf{b}, \qquad \text{in } \Omega, \tag{6.1a}$$

$$\mathcal{B}\mathbf{u} = \alpha\mathbf{f}, \qquad \text{on } \Gamma. \tag{6.1b}$$

A. Luongo et al., *Perturbation Methods and Nonlinear Phenomena*,
Synthesis Lectures on Engineering, Science, and Technology,
https://doi.org/10.1007/978-3-031-49397-3_6

Here, the Eqs. 5.2 have been used and a load multiplier α has been considered to affect the loads $\mathbf{b}, \mathbf{f}$. It should be stressed that $\mathcal{K}$ is a *non-singular operator* (in the sense that the associated homogeneous problem $\mathcal{K}\mathbf{u} = \mathbf{0}$, with the boundary conditions $\mathcal{B}\mathbf{u} = \mathbf{0}$, admits only the trivial solution $\mathbf{u} = \mathbf{0}$), i.e., any mispositioning of the constraints is excluded.

Goal of the analysis is to build-up the curve $\mathbf{u} = \mathbf{u}(\mathbf{s}; \alpha)$, which describes the static nonlinear response of the structure. To this end, a perturbation parameter $0 < \epsilon \ll 1$ is artificially introduced in the equation, via the rescaling $\mathbf{u} \to \epsilon\hat{\mathbf{u}}$, $\alpha \to \epsilon\hat{\alpha}$, with $\hat{\mathbf{u}} = O(1)$, $\hat{\alpha} = O(1)$.[1] Since $\boldsymbol{n}_h(\epsilon\mathbf{u}, \epsilon\mathbf{u}, \ldots) = \epsilon^h \boldsymbol{n}_h(\mathbf{u}, \mathbf{u}, \ldots)$, being the function homogeneous, after division by ϵ, the Eqs. 6.1 read (hat omitted):

$$\mathcal{K}\mathbf{u} = -\epsilon \boldsymbol{n}_2(\mathbf{u}, \mathbf{u}) - \epsilon^2 \boldsymbol{n}_3(\mathbf{u}, \mathbf{u}, \mathbf{u}) + \cdots + \alpha\mathbf{b}, \tag{6.2a}$$

$$\mathcal{B}\mathbf{u} = \alpha\mathbf{f}. \tag{6.2b}$$

The unknown static response $\mathbf{u} = \mathbf{u}\left(\mathbf{s}; \epsilon\hat{\alpha}\right) = \mathbf{u}(\mathbf{s}; \epsilon)$ is expanded in MacLaurin series, as:

$$\mathbf{u} = \mathbf{u}_0 + \epsilon\mathbf{u}_1 + \epsilon^2\mathbf{u}_2 + \cdots, \tag{6.3}$$

where the $\mathbf{u}_h(\mathbf{s})$ vectors are unknown. By substituting the series in Eqs. 6.2, and requiring that they are satisfied for any ϵ, the terms with the same power of ϵ must be vanished separately. This generates a set of *perturbation equations*, namely[2]:

Order ϵ^0:

$$\mathcal{K}\mathbf{u}_0 = \alpha\mathbf{b}, \tag{6.4a}$$

$$\mathcal{B}\mathbf{u}_0 = \alpha\mathbf{f}. \tag{6.4b}$$

[1] The multiplier α itself could be used as perturbation parameter, but the procedure illustrated here is preferred, since systematic.

[2] It has been exploited, that $\boldsymbol{n}_2(\mathbf{u}+\mathbf{v}, \mathbf{u}+\mathbf{v}) = \boldsymbol{n}_2(\mathbf{u}, \mathbf{u}) + 2\boldsymbol{n}_2(\mathbf{u}, \mathbf{v}) + \boldsymbol{n}_2(\mathbf{v}, \mathbf{v})$, where $\boldsymbol{n}_2(\mathbf{u}, \mathbf{v}) = \boldsymbol{n}_2(\mathbf{v}, \mathbf{u})$ (see also Note 3 in the Appendix B). As a scalar example, if $n_2(u, u) := uu'$, it follows that $n_2(u+v, u+v) = (u+v)(u+v)' = uu' + (u'v + uv') + vv'$, from which:

$$n_2(u, v) = \frac{1}{2}(u'v + uv').$$

Similarly, for cubic operators, it is $\boldsymbol{n}_3(\mathbf{u}+\mathbf{v}, \mathbf{u}+\mathbf{v}, \mathbf{u}+\mathbf{v}) = \boldsymbol{n}_3(\mathbf{u}, \mathbf{u}, \mathbf{u}) + 3\boldsymbol{n}_3(\mathbf{u}, \mathbf{u}, \mathbf{v}) + 3\boldsymbol{n}_3(\mathbf{u}, \mathbf{v}, \mathbf{v}) + \boldsymbol{n}_3(\mathbf{v}, \mathbf{v}, \mathbf{v})$. For example, if $n_3(u, u, u) =: uu'u''$, by the same reasoning, it follows:

$$n_3(u, u, v) = \frac{1}{3}(uu'v'' + uu''v' + u'u''v),$$

$$n_3(u, v, v) = \frac{1}{3}(uv'v'' + u'vv'' + u''vv').$$

Order ϵ^1:

$$\mathcal{K}\mathbf{u}_1 = -\boldsymbol{n}_2(\mathbf{u}_0, \mathbf{u}_0), \tag{6.5a}$$

$$\mathcal{B}\mathbf{u}_1 = \mathbf{0}. \tag{6.5b}$$

Order ϵ^2:

$$\mathcal{K}\mathbf{u}_2 = -2\boldsymbol{n}_2(\mathbf{u}_0, \mathbf{u}_1) - \boldsymbol{n}_3(\mathbf{u}_0, \mathbf{u}_0, \mathbf{u}_0), \tag{6.6a}$$

$$\mathcal{B}\mathbf{u}_2 = \mathbf{0}. \tag{6.6b}$$

The perturbation equations possess the following properties (common to most perturbation methods): (a) they are linear in $\mathbf{u}_h,\ h = 0, 1, 2, \ldots$; (b) they contain terms up to $\mathbf{u}_{h-1}$ on the right hand side; (c) they are governed by the same operator $\mathcal{K}$ in the domain, and $\mathcal{B}$ on the boundary. Consequently, they can be solved in sequence to furnish *unique solutions*, of the type $\mathbf{u}_h = \alpha^{h+1}\boldsymbol{\chi}_h(\mathbf{s})$. By substituting them in the series Eq. 6.3, and reabsorbing the ϵ parameter (i.e., coming back to the unrescaled quantities), we finally obtain:

$$\mathbf{u} = \alpha\boldsymbol{\chi}_0(\mathbf{s}) + \alpha^2\boldsymbol{\chi}_1(\mathbf{s}) + \alpha^3\boldsymbol{\chi}_2(\mathbf{s}) + \cdots. \tag{6.7}$$

In this expression, the first contribution on the right hand side represents the linear elastostatic response of the structure; the successive ones correct the linear solution, accounting for quadratic, cubic, ..., nonlinearities. Worked out examples of elastostatics will be presented in the Problems 7.1 and 7.2.

6.3 Buckling and Postbuckling

The buckling problem, already tackled in the Sect. 2.3 with reference to the sample system, is now discussed for a general system. The strained parameter method, in static field, is used [10, 12, 13, 15–17].

A key point of the perturbation method used in Sect. 6.2 is that $\mathcal{K}$ is a non-singular operator. However, when a buckling problem is addressed, the interest just relies in analyzing the neighborhood of the bifurcation point, in the parameter space, at which $\mathcal{K}$ is singular. As an example, by referring to the Euler beam, the response of the beam is sought for values of the axial load slightly larger than the critical one. In this class of problems (of static bifurcations), the method of the Sect. 6.2 must be modified to overcome some algorithmic difficulties, as explained soon.

Perturbation Equations

The static equilibrium equations of a continuous system, presolicited by forces proportional to a load multiplier μ, are deduced by Eqs. 5.3, 5.2, ignoring the dynamic effects and taking $\mathbf{u} = \mathbf{u}(\mathbf{s})$, namely:

$$\left(\mathcal{K}_e + \mu \mathcal{K}_g\right) \mathbf{u} = -\boldsymbol{n}_2 (\mathbf{u}, \mathbf{u}; \mu) - \boldsymbol{n}_3 (\mathbf{u}, \mathbf{u}, \mathbf{u}; \mu) + \cdots, \qquad \text{in } \Omega, \tag{6.8a}$$

$$\left(\mathcal{B}_e + \mu \mathcal{B}_g\right) \mathbf{u} = \mathbf{0}, \qquad \text{on } \Gamma. \tag{6.8b}$$

It is assumed that $\boldsymbol{n}_h (\mathbf{0}, \mathbf{0}, \ldots; \mu) = \mathbf{0}$ for any h, so that the system admits the trivial fundamental path $\mathbf{u} = \mathbf{0},\ \forall \mu$; the inextensible Euler beam falls in this category.[3] The task is to find the *bifurcated path* $\mathbf{u} = \mathbf{u}(\mathbf{s}; \mu)$ (or, in parametric form, $\mathbf{u} = \mathbf{u}(\mathbf{s}; \epsilon)$, $\mu = \mu(\epsilon)$) which branches off from the trivial path at the (unknown) bifurcation point $(\mu, \mathbf{u}) = (\mu_c, \mathbf{0})$. To this end, the rescaling $\mathbf{u} \to \epsilon \hat{\mathbf{u}}$ is performed, in order a perturbation parameter appears; then, *both* the displacement $\mathbf{u}$ *and* the load μ are expanded in series as follows:

$$\hat{\mathbf{u}} = \mathbf{u}_0 + \epsilon \mathbf{u}_1 + \epsilon^2 \mathbf{u}_2 + \cdots, \tag{6.9a}$$

$$\mu = \mu_0 + \epsilon \mu_1 + \epsilon^2 \mu_2 + \cdots, \tag{6.9b}$$

whose coefficients $\mathbf{u}_h (\mathbf{s})$ and μ_h are all unknown. By substituting the series Eq. 6.9 in Eqs. 6.8, and separately zeroing the terms with the same powers of ϵ, the following perturbation equations are derived[4]:

Order ϵ^0:

$$\left(\mathcal{K}_e + \mu_0 \mathcal{K}_g\right) \mathbf{u}_0 = \mathbf{0}, \tag{6.10a}$$

$$\left(\mathcal{B}_e + \mu_0 \mathcal{B}_g\right) \mathbf{u}_0 = \mathbf{0}. \tag{6.10b}$$

Order ϵ^1:

$$\left(\mathcal{K}_e + \mu_0 \mathcal{K}_g\right) \mathbf{u}_1 = -\mu_1 \mathcal{K}_g \mathbf{u}_0 - \boldsymbol{n}_2 (\mathbf{u}_0, \mathbf{u}_0; \mu_0), \tag{6.11a}$$

$$\left(\mathcal{B}_e + \mu_0 \mathcal{B}_g\right) \mathbf{u}_1 = -\mu_1 \mathcal{B}_g \mathbf{u}_0. \tag{6.11b}$$

[3] If this is not the case, a change of variable $\mathbf{u} = \mathbf{u}^f(\mathbf{s}; \mu) + \mathbf{v}(\mathbf{s})$, with $\mathbf{u}^f$ the known *non-trivial path*, permits to reduce the problem to the simpler case, although the dependence on μ of the total stiffness matrix $\mathcal{K}$ is no longer linear (see, e.g., [10]).

[4] The vector $\boldsymbol{n}_2$ has been expanded with respect to the load as $\boldsymbol{n}_2(\mathbf{u}, \mathbf{u}; \mu) = \boldsymbol{n}_2(\mathbf{u}, \mathbf{u}; \mu_0) + \epsilon \mu_1 \boldsymbol{n}_{2,\mu}(\mathbf{u}, \mathbf{u}; \mu_0) + \cdots$, where $\boldsymbol{n}_{2,\mu} := \frac{\partial \boldsymbol{n}_2}{\partial \mu}$.

Order ϵ^2:

$$\left(\mathcal{K}_e + \mu_0 \mathcal{K}_g\right) \mathbf{u}_2 = -\mu_2 \mathcal{K}_g \mathbf{u}_0 - \mu_1 \mathcal{K}_g \mathbf{u}_1 - 2\boldsymbol{n}_2 \left(\mathbf{u}_0, \mathbf{u}_1; \mu_0\right) - \mu_1 \boldsymbol{n}_{2,\mu} \left(\mathbf{u}_0, \mathbf{u}_0; \mu_0\right) - \boldsymbol{n}_3 \left(\mathbf{u}_0, \mathbf{u}_0, \mathbf{u}_0; \mu_0\right), \tag{6.12a}$$

$$\left(\mathcal{B}_e + \mu_0 \mathcal{B}_g\right) \mathbf{u}_2 = -\mu_2 \mathcal{B}_g \mathbf{u}_0 - \mu_1 \mathcal{B}_g \mathbf{u}_1. \tag{6.12b}$$

Differently from the static analysis, now the generating ϵ^0 order Eq. 6.10 is homogeneous, while the successive ones are non-homogeneous. In order to obtain non-trivial solution from Eq. 6.10, μ_0 must assume specific values (i.e., it must be an eigenvalue of the boundary value problem). Hence, the total stiffness operator becomes singular. As a consequence, the successive non-homogeneous perturbation Eqs. 6.11, 6.12, …cannot be solved, unless the $\mu_1, \mu_2, \ldots$parameters take special values, allowing *solvability* of the singular problems. These conditions provide the coefficients of the series expansion of the load.

Another consequence of the singularity of the operator is that its solutions at the different orders are not unique, as they are defined to within an indeterminate parameter.[5] To remove the indeterminacy, an arbitrary normalization condition is introduced, e.g., $u_i\left(\mathbf{s}^*; \epsilon\right) = a$ where u_i is the ith component of $\mathbf{u}$, $\mathbf{s}^*$ is a properly selected point, and a an assigned amplitude. Accounting for the series expansion Eq. 6.9a, a set of normalization conditions follow at each order, namely: $u_{i0}\left(\mathbf{s}^*\right) = a$, $u_{i1}\left(\mathbf{s}^*\right) = 0$, $u_{i2}\left(\mathbf{s}^*\right) = 0, \ldots$.

Solution

The algorithm develops as follows. Equation 6.10 admits an infinite number of eigenvalues (here assumed all distinct) and eigenvectors. By taking the smallest eigenvalue,[6] $\mu_0 =: \mu_c$ is found as the (first) *critical load*. The associated normalized eigenvector is the critical mode $\mathbf{u}_0 =: a\boldsymbol{\phi}_c$, with $\phi_{ic}\left(\mathbf{s}^*\right) = 1$. With these results, the ϵ order perturbation equation reads:

$$\left(\mathcal{K}_e + \mu_c \mathcal{K}_g\right) \mathbf{u}_1 = -a\mu_1 \mathcal{K}_g \boldsymbol{\phi}_c - a^2 \boldsymbol{n}_2 \left(\boldsymbol{\phi}_c, \boldsymbol{\phi}_c; \mu_c\right), \tag{6.13a}$$

$$\left(\mathcal{B}_e + \mu_c \mathcal{B}_g\right) \mathbf{u}_1 = -a\mu_1 \mathcal{B}_g \boldsymbol{\phi}_c. \tag{6.13b}$$

Since this problem is singular, in order it to admit non-trivial solutions, the know terms, both in the field and at the boundary, must spend zero virtual work on the critical mode (see the Appendix A, Sect. A.3 for a more detailed discussion on this topic), i.e.:

$$\int_\Omega \boldsymbol{\phi}_c^T \left[a\mu_1 \mathcal{K}_g \boldsymbol{\phi}_c + a^2 \boldsymbol{n}_2 \left(\boldsymbol{\phi}_c, \boldsymbol{\phi}_c; \mu_c\right)\right] \mathrm{d}\Omega + \int_\Gamma \boldsymbol{\phi}_c^T \left[a\mu_1 \mathcal{B}_g \boldsymbol{\phi}_c\right] \mathrm{d}\Gamma = 0 \tag{6.14}$$

[5] It is assumed here that the operator possesses a one-dimensional kernel.

[6] Usually, the interest is on $\mu > 0$, for example, on compression loads acting on the Euler beam.

must hold. This is a linear equation, from which $\mu_1 = C_1 a$ is drawn, with C_1 a constant defined as follows:

$$C_1 := -\frac{\int_\Omega \boldsymbol{\phi}_c^T \boldsymbol{n}_2\left(\boldsymbol{\phi}_c, \boldsymbol{\phi}_c; \mu_c\right) \mathrm{d}\Omega}{\int_\Omega \boldsymbol{\phi}_c^T \mathcal{K}_g \boldsymbol{\phi}_c \mathrm{d}\Omega + \int_\Gamma \boldsymbol{\phi}_c^T \mathcal{B}_g \boldsymbol{\phi}_c \mathrm{d}\Gamma}. \tag{6.15}$$

If the expansion is truncated at this order, Eqs. 6.9 provide the tangent at the bifurcated path, i.e., $\mathbf{u} =: a\boldsymbol{\phi}_c$, $\mu = \mu_c + \epsilon C_1 a$. It should be noticed that this low-order solution does not account for the modification of the buckling pattern, which is instead frozen at the buckling mode. Therefore, the solution coincides with that would be obtained by applying the Galerkin method to the structure, i.e., after having reduced it to a single DOF system.

To improve the solution, a further step is needed. First, Eq. 6.13 must be solved, to get $\mathbf{u}_1 = a^2 \boldsymbol{\chi}_1(\mathbf{s})$, where $\boldsymbol{\chi}_1(\mathbf{s})$ is rendered unique by exploiting the normalization condition $\chi_{1i}(\mathbf{s}^*) = 0$. By substituting all these results in the ϵ^2 order perturbation Eq. 6.12, this latter reads:

$$\begin{aligned}\left(\mathcal{K}_e + \mu_0 \mathcal{K}_g\right) \mathbf{u}_2 &= -a\mu_2 \mathcal{K}_g \boldsymbol{\phi}_c - a^3 C_1 \mathcal{K}_g \boldsymbol{\chi}_1 - 2a^3 \boldsymbol{n}_2\left(\boldsymbol{\phi}_c, \boldsymbol{\chi}_1; \mu_c\right) \\ &\quad - a^3 C_1 \boldsymbol{n}_{2,\mu}\left(\boldsymbol{\phi}_c, \boldsymbol{\phi}_c; \mu_c\right) - a^3 \boldsymbol{n}_3\left(\boldsymbol{\phi}_c, \boldsymbol{\phi}_c, \boldsymbol{\phi}_c; \mu_c\right),\end{aligned} \tag{6.16a}$$

$$\left(\mathcal{B}_e + \mu_0 \mathcal{B}_g\right) \mathbf{u}_2 = -a\mu_2 \mathcal{B}_g \boldsymbol{\phi}_c - a^3 C_1 \mathcal{B}_g \boldsymbol{\chi}_1. \tag{6.16b}$$

Solvability of the singular Eqs. 6.16 requires that:

$$\begin{aligned}&\int_\Omega \boldsymbol{\phi}_c^T \left\{a\mu_2 \mathcal{K}_g \boldsymbol{\phi}_c + a^3 \left[C_1 \mathcal{K}_g \boldsymbol{\chi}_1 + 2\boldsymbol{n}_2\left(\boldsymbol{\phi}_c, \boldsymbol{\chi}_1; \mu_c\right) + C_1 \boldsymbol{n}_{2,\mu}\left(\boldsymbol{\phi}_c, \boldsymbol{\phi}_c; \mu_c\right)\right.\right. \\ &\quad \left.\left. + \boldsymbol{n}_3\left(\boldsymbol{\phi}_c, \boldsymbol{\phi}_c, \boldsymbol{\phi}_c; \mu_c\right)\right]\right\} \mathrm{d}\Omega + \int_\Gamma \boldsymbol{\phi}_c^T \left[a\mu_2 \mathcal{B}_g \boldsymbol{\phi}_c + a^3 C_1 \mathcal{B}_g \boldsymbol{\chi}_1\right] \mathrm{d}\Gamma = 0,\end{aligned} \tag{6.17}$$

whose solution supplies $\mu_2 = C_2 a^2$, where:

$$\begin{aligned}C_2 := -&\frac{1}{\int_\Omega \boldsymbol{\phi}_c^T \mathcal{K}_g \boldsymbol{\phi}_c \mathrm{d}\Omega + \int_\Gamma \boldsymbol{\phi}_c^T \mathcal{B}_g \boldsymbol{\phi}_c \mathrm{d}\Gamma} \left\{\int_\Omega \boldsymbol{\phi}_c^T \Big[C_1 \mathcal{K}_g \boldsymbol{\chi}_1 + 2\boldsymbol{n}_2\left(\boldsymbol{\phi}_c, \boldsymbol{\chi}_1; \mu_c\right)\right. \\ &\left. + C_1 \boldsymbol{n}_{2,\mu}\left(\boldsymbol{\phi}_c, \boldsymbol{\phi}_c; \mu_c\right) + \boldsymbol{n}_3\left(\boldsymbol{\phi}_c, \boldsymbol{\phi}_c, \boldsymbol{\phi}_c; \mu_c\right)\Big] \mathrm{d}\Omega + \int_\Gamma C_1 \boldsymbol{\phi}_c^T \mathcal{B}_g \boldsymbol{\chi}_1 \mathrm{d}\Gamma\right\}.\end{aligned} \tag{6.18}$$

By summarizing, at the ϵ^2 order, the bifurcated path is expressed by:

$$\mathbf{u} = a\boldsymbol{\phi}_c + a^2 \boldsymbol{\chi}_1 + \cdots, \tag{6.19a}$$

$$\mu = \mu_c + aC_1 + a^2 C_2 + \cdots, \tag{6.19b}$$

in which the perturbation parameter has been reabsorbed. This equation gives, for any assigned displacement $a \equiv u_i(\mathbf{s}^*)$, the associated displacement field $\mathbf{u}(\mathbf{s})$ and the load μ. This latter is a parabolic function of the amplitude, which captures the exact tangent and curvature at the bifurcation ($a = 0$) and extrapolates the solution to $a \neq 0$. It is important to stress that $a^2 \chi_1(\mathbf{s})$ describes the modification of the buckling pattern with the amplitude, and that C_2 accounts for it.[7] An example of buckling problem is addressed in the Problem 7.3.

6.4 External Resonance

The external resonance phenomenon has been illustrated in the Sects. 3.2 and 3.4 referring to the sample system. Here, the problem is reconsidered for a general structure. The Multiple Scale Method (MSM) is systematically used [1, 6–8, 11, 12, 14].

When the external forces are time-varying, the dynamic effects due to the mass and damping must be included in the model. In particular, when the excitation is harmonic of frequency Ω, close to a natural frequency ω_r, the *primary external resonance* phenomenon takes place; when Ω is a multiple or sub-multiple of ω_r, the *sub- or super-harmonic resonance*, respectively, occurs.

The problem is governed by Eqs. 5.1, in which the time-dependence of the field and boundary forces is made explicit:

$$\mathcal{M}\ddot{\mathbf{u}} + \mathcal{C}\dot{\mathbf{u}} + \mathcal{K}\mathbf{u} + \boldsymbol{n}(\mathbf{u}) = \alpha \mathbf{b}_0(\mathbf{s}) \cos(\Omega t), \quad \text{in } \Omega, \tag{6.20a}$$

$$\mathcal{D}\dot{\mathbf{u}} + \mathcal{B}\mathbf{u} = \alpha \mathbf{f}_0(\mathbf{s}) \cos(\Omega t), \quad \text{on } \Gamma, \tag{6.20b}$$

where $\boldsymbol{n}(\mathbf{u}) = \boldsymbol{n}_2(\mathbf{u}, \mathbf{u}) + \boldsymbol{n}_3(\mathbf{u}, \mathbf{u}, \mathbf{u}) + \cdots$ and α is a load multiplier. Here $\mathcal{K} \equiv \mathcal{K}_e$ is the elastic stiffness operator, which is self-adjoint together with its boundary conditions, due to the existence of the elastic energy.[8]

[7] The structure, therefore, behaves as a single degree of freedom system, but its configurations does not span a subspace (i.e., the kernel of the stiffness operator), but rather a manifold, having the same dimension and tangent to it. The analogy with the Center Manifold Theory [4, 5, 18, 19] (mentioned in the Remark 4.3), relevant to dynamical systems, should be noticed.

[8] An operator $\mathcal{L}$, with its boundary conditions $\mathcal{B}$, is self-adjoint when, for any vectors $\boldsymbol{\phi}$, $\boldsymbol{\psi}$ defined in the function space on which $\mathcal{L}$ and $\mathcal{B}$ operate, the following *Extended Green Identity* holds:

$$\int_\Omega \boldsymbol{\psi}^T \mathcal{L}\boldsymbol{\phi}\, d\Omega + \int_\Gamma \boldsymbol{\psi}^T \mathcal{B}\boldsymbol{\phi}\, d\Gamma = \int_\Omega \boldsymbol{\phi}^T \mathcal{L}\boldsymbol{\psi}\, d\Omega + \int_\Gamma \boldsymbol{\phi}^T \mathcal{B}\boldsymbol{\psi}\, d\Gamma.$$

Modal Properties

First, the linear free and undamped system is considered, which reads:

$$\mathcal{M}\ddot{\mathbf{u}} + \mathcal{K}\mathbf{u} = \mathbf{0}, \tag{6.21a}$$

$$\mathcal{B}\mathbf{u} = \mathbf{0}. \tag{6.21b}$$

By letting $\mathbf{u} = \boldsymbol{\phi}_k(\mathbf{s})\, e^{i\omega_k t}$, a spatial boundary value problem follows:

$$\left(\mathcal{K} - \omega_k^2 \mathcal{M}\right) \boldsymbol{\phi}_k = \mathbf{0}, \tag{6.22a}$$

$$\mathcal{B}\boldsymbol{\phi}_k = \mathbf{0}, \tag{6.22b}$$

whose eigenpairs $\left(\omega_k, \boldsymbol{\phi}_k\right)$, $k = 1, 2, \ldots$, are the natural frequencies and modes, respectively. Due to the self-adjointness of $\mathcal{K}$ ($\mathcal{M}$ being always self-adjoint), frequencies and modes are real; moreover, the modes are mutually orthogonal.[9]

6.4.1 Primary Resonance

The primary resonance case, $\Omega \simeq \omega_r$, is addressed first. To tackle Eqs. 6.20 by the MSM, the displacement is first rescaled as $\mathbf{u} \to \epsilon\hat{\mathbf{u}}$; moreover, the external load parameter is taken as $\alpha \to \epsilon^3\hat{\alpha}$ (*soft excitation*) and damping also rescaled as $(\mathcal{C}, \mathcal{D}) \to \epsilon^2\left(\hat{\mathcal{C}}, \hat{\mathcal{D}}\right)$,[10] finally, the closeness of the external frequency to the rth natural frequency is expressed as $\Omega = \omega_r + \epsilon^2\hat{\sigma}_r$. The Eqs. 6.20, when rescaled, become (hat omitted):

$$\mathcal{M}\ddot{\mathbf{u}} + \mathcal{K}\mathbf{u} = -\epsilon \boldsymbol{n}_2(\mathbf{u}, \mathbf{u}) - \epsilon^2\left[\mathcal{C}\dot{\mathbf{u}} + \boldsymbol{n}_3(\mathbf{u}, \mathbf{u}, \mathbf{u}) + \left(\frac{1}{2}\alpha\mathbf{b}_0 e^{i(\omega_r + \epsilon^2\sigma_r)t} + \text{c.c.}\right)\right], \tag{6.23a}$$

$$\mathcal{B}\mathbf{u} = -\epsilon^2\left[\mathcal{D}\dot{\mathbf{u}} + \left(\frac{1}{2}\alpha\mathbf{f}_0 e^{i(\omega_r + \epsilon^2\sigma_r)t} + \text{c.c.}\right)\right], \tag{6.23b}$$

where the acronym c.c. denotes the complex conjugates of the preceding terms.

---

[9] Namely, for any h, k:

$$\int_\Omega \boldsymbol{\phi}_h^T \mathcal{K}\boldsymbol{\phi}_k \, \mathrm{d}\Omega + \int_\Gamma \boldsymbol{\phi}_h^T \mathcal{B}\boldsymbol{\phi}_k \, \mathrm{d}\Gamma = \delta_{hk}\omega_k^2,$$

where δ_{hk} is the Kronecker operator, and the (arbitrary) normalization $\int_\Omega \boldsymbol{\phi}_k^T \mathcal{M}\boldsymbol{\phi}_k \, \mathrm{d}\Omega = 1$ has been used.

[10] Rescaling is performed in such a way the load and the damping forces are made of the same order of $|\mathbf{u}|^3$, for the reasons which will appear clear soon.

Multiple Scale Perturbation Equations

The MSM is used [12], according to which $\mathbf{u}(s, t)$ depends on several independent time-scales:

$$t_0 := t, \quad t_1 := \epsilon t, \quad t_2 := \epsilon^2 t, \quad \ldots \tag{6.24}$$

i.e., $\mathbf{u}(\mathbf{s}, t) = \mathbf{u}(\mathbf{s}, t_0(t), t_1(t), t_2(t)), \ldots)$. By using the chain rule, it follows:

$$\dot{\mathbf{u}} = \left(\partial_0 + \epsilon \partial_1 + \epsilon^2 \partial_2 + \cdots\right) \mathbf{u}, \tag{6.25a}$$

$$\ddot{\mathbf{u}} = \left(\partial_0 + \epsilon \partial_1 + \epsilon^2 \partial_2 + \cdots\right)^2 \mathbf{u}, \tag{6.25b}$$

where $\partial_k := \frac{\partial}{\partial t_k}$. By expanding $\mathbf{u}$ is series of ϵ, as $\mathbf{u} = \mathbf{u}_0 + \epsilon \mathbf{u}_1 + \epsilon^2 \mathbf{u}_2 + \cdots$, the perturbation equations are finally derived; they read:

Order ϵ^0:

$$\mathcal{M}\partial_0^2 \mathbf{u}_0 + \mathcal{K}\mathbf{u}_0 = \mathbf{0}, \tag{6.26a}$$

$$\mathcal{B}\mathbf{u}_0 = \mathbf{0}. \tag{6.26b}$$

Order ϵ^1:

$$\mathcal{M}\partial_0^2 \mathbf{u}_1 + \mathcal{K}\mathbf{u}_1 = -2\mathcal{M}\partial_0\partial_1 \mathbf{u}_0 - \boldsymbol{n}_2\left(\mathbf{u}_0, \mathbf{u}_0\right), \tag{6.27a}$$

$$\mathcal{B}\mathbf{u}_1 = \mathbf{0}. \tag{6.27b}$$

Order ϵ^2:

$$\begin{aligned} \mathcal{M}\partial_0^2 \mathbf{u}_2 + \mathcal{K}\mathbf{u}_2 = &-\mathcal{M}\partial_1^2 \mathbf{u}_0 - 2\mathcal{M}\partial_0\partial_2 \mathbf{u}_0 - 2\mathcal{M}\partial_0\partial_1 \mathbf{u}_1 - \mathcal{C}\partial_0 \mathbf{u}_0 \\ &- 2\boldsymbol{n}_2\left(\mathbf{u}_0, \mathbf{u}_1\right) - \boldsymbol{n}_3\left(\mathbf{u}_0, \mathbf{u}_0, \mathbf{u}_0\right) \\ &+ \left(\frac{1}{2}\alpha \mathbf{b}_0 e^{i\sigma_r t_2} e^{i\omega_r t_0} + \text{c.c.}\right), \end{aligned} \tag{6.28a}$$

$$\mathcal{B}\mathbf{u}_2 = -\mathcal{D}\partial_0 \mathbf{u}_0 + \left(\frac{1}{2}\alpha \mathbf{f}_0 e^{i\sigma_r t_2} e^{i\omega_r t_0} + \text{c.c.}\right). \tag{6.28b}$$

Solution

Equations 6.26–6.28 are now solved in sequence. Since the modes different from the rth are not in resonance[11] and are damped, a monomodal solution is taken to Eq. 6.26:

$$\mathbf{u}_0 = A_r\left(t_1, t_2, \ldots\right) \boldsymbol{\phi}_r\left(\mathbf{s}\right) e^{i\omega_r t_0} + \text{c.c.}, \tag{6.29}$$

where $A_r := \frac{1}{2} a_r\left(t_1, t_2, \ldots\right) e^{i\varphi_r\left(t_1, t_2, \ldots\right)}$ is an unknown complex amplitude, accounting for modulation of the real amplitude a_r and phase φ_r on the slow time scales. With this solution, the ϵ order perturbation Eqs. 6.27 become[12]:

$$\begin{aligned}\mathcal{M}\partial_0^2\mathbf{u}_1 + \mathcal{K}\mathbf{u}_1 &= -2i\omega_r\partial_1 A_r \mathcal{M}\boldsymbol{\phi}_r e^{i\omega_r t_0} \\ &\quad - \left(A_r^2 e^{2i\omega_r t_0} + A_r\bar{A}_r\right) \boldsymbol{n}_2\left(\boldsymbol{\phi}_r, \boldsymbol{\phi}_r\right) + \text{c.c.},\end{aligned} \tag{6.30a}$$

$$\mathcal{B}\mathbf{u}_1 = \mathbf{0}, \tag{6.30b}$$

where an overbar denotes complex conjugate. On its right hand side, three frequencies appear at exponential, i.e., 0, ω_r, $2\omega_r$ (and, of course, $-\omega_r$, $-2\omega_r$). If $\omega_k \neq 2\omega_r$ $\forall k$, the only resonant frequency is ω_r, which has to be made orthogonal to $\boldsymbol{\phi}_r$ to avoid that the solution diverges on the t_0 scale (compatibility condition).[13] From this condition, the equation $\partial_1 A_r = 0$ follows, i.e., the complex amplitude does not depend on t_1. Therefore, a further step must be performed, in order to find its dependence on t_2.

---

[11] Internal resonances [11] are excluded, i.e., the natural frequencies are assumed to be incommensurable.

[12] If $\mathbf{x} = \mathbf{a} + \bar{\mathbf{a}}$, $\mathbf{y} = \mathbf{b} + \bar{\mathbf{b}}$, then $\boldsymbol{n}_2\left(\mathbf{x}, \mathbf{x}\right) = \boldsymbol{n}_2\left(\mathbf{a}, \mathbf{a}\right) + 2\boldsymbol{n}_2\left(\mathbf{a}, \bar{\mathbf{a}}\right) + \boldsymbol{n}_2\left(\bar{\mathbf{a}}, \bar{\mathbf{a}}\right) = \boldsymbol{n}_2\left(\mathbf{a}, \mathbf{a}\right) + \boldsymbol{n}_2\left(\mathbf{a}, \bar{\mathbf{a}}\right) + \text{c.c.}$, the overbar indicating complex conjugate. Similarly, $\boldsymbol{n}_2\left(\mathbf{x}, \mathbf{y}\right) = \boldsymbol{n}_2\left(\mathbf{a}, \mathbf{b}\right) + \boldsymbol{n}_2\left(\mathbf{a}, \bar{\mathbf{b}}\right) + \text{c.c.}$ and $\boldsymbol{n}_3\left(\mathbf{x}, \mathbf{x}, \mathbf{x}\right) = \boldsymbol{n}_3\left(\mathbf{a}, \mathbf{a}, \mathbf{a}\right) + 3\boldsymbol{n}_3\left(\mathbf{a}, \mathbf{a}, \bar{\mathbf{a}}\right) + \text{c.c.}$, to be used later.

[13] Indeed, considered the problem:

$$\begin{aligned}\mathcal{M}\ddot{\mathbf{u}} + \mathcal{K}\mathbf{u} &= \mathbf{p}e^{i\omega_r t}, &&\text{in } \Omega, \\ \mathcal{B}\mathbf{u} &= \mathbf{q}e^{i\omega_r t}, &&\text{on } \Gamma,\end{aligned}$$

and taken the solution as $\mathbf{u} = \hat{\mathbf{u}}e^{i\omega_r t}$, a space-differential problem follows:

$$\begin{aligned}\left(\mathcal{K} - \omega_r^2\mathcal{M}\right)\hat{\mathbf{u}} &= \mathbf{p}, \\ \mathcal{B}\hat{\mathbf{u}} &= \mathbf{q}.\end{aligned}$$

Since the operator $\mathcal{K} - \omega_r^2\mathcal{M}$ is singular, whose kernel is spanned by $\boldsymbol{\phi}_r$, the known terms $\mathbf{p}, \mathbf{q}$ must be orthogonal to $\boldsymbol{\phi}_r$ (i.e., the forces must spend zero virtual work on the 'floppy' mode admitted by the operator), namely:

$$\int_\Omega \boldsymbol{\phi}_r^T\mathbf{p}\,\mathrm{d}\Omega + \int_\Gamma \boldsymbol{\phi}_r^T\mathbf{q}\,\mathrm{d}\Gamma = 0.$$

By exploiting the linearity, the solution to the perturbation Eq. 6.30 is put in the form:

$$\mathbf{u}_1 = A_r^2 \boldsymbol{\chi}_{12}(\mathbf{s}) e^{2i\omega_r t_0} + A_r \bar{A}_r \boldsymbol{\chi}_{10}(\mathbf{s}) + \text{c.c.}, \tag{6.31}$$

where the $\boldsymbol{\chi}$ ($\mathbf{s}$) functions are uniquely determined by solving non-homogeneous boundary value problems governed by the non-singular operators $\mathcal{K} - 4\omega_r^2 \mathcal{M}$ and $\mathcal{K}$, respectively. They describe the modification of the pattern of the response, which is therefore made of a linear part $\mathbf{u}_0$, equal to the rth natural mode, and a nonlinear part $\mathbf{u}_1$, proportional to the squared amplitude. The complementary solution is ignored, as a normalization condition.

With these results, the perturbation Eqs. 6.28 become:

$$\mathcal{M}\partial_0^2 \mathbf{u}_2 + \mathcal{K}\mathbf{u}_2 = \mathbf{p}_{21} e^{i\omega_r t_0} + \mathbf{p}_{22} e^{2i\omega_r t_0} + \mathbf{p}_{23} e^{3i\omega_r t_0} + \text{c.c.}, \tag{6.32a}$$

$$\mathcal{B}\mathbf{u}_2 = \mathbf{q}_{21} e^{i\omega_r t_0} + \text{c.c.}, \tag{6.32b}$$

where:

$$\begin{aligned}
\mathbf{p}_{21} :=& - A_r^2 \bar{A}_r [2\boldsymbol{n}_2 (\boldsymbol{\phi}_r, \boldsymbol{\chi}_{10}) + 2\boldsymbol{n}_2 (\boldsymbol{\phi}_r, \boldsymbol{\chi}_{12}) + 2\boldsymbol{n}_2 (\boldsymbol{\phi}_r, \bar{\boldsymbol{\chi}}_{10}) \\
&+ 3\boldsymbol{n}_3 (\boldsymbol{\phi}_r, \boldsymbol{\phi}_r, \boldsymbol{\phi}_r)] - i\omega_r A_r \mathcal{C}\boldsymbol{\phi}_r \\
&- 2i\omega_r \partial_2 A_r \mathcal{M}\boldsymbol{\phi}_r - \partial_1^2 A_r \mathcal{M}\boldsymbol{\phi}_r + \frac{1}{2}\alpha \mathbf{b}_0 e^{i\sigma_r t_2},
\end{aligned} \tag{6.33a}$$

$$\mathbf{p}_{22} := - 8i\omega_r A_r \partial_1 A_r \mathcal{M}\boldsymbol{\chi}_{12}, \tag{6.33b}$$

$$\mathbf{p}_{23} := - A_r^3 [2\boldsymbol{n}_2 (\boldsymbol{\phi}_r, \boldsymbol{\chi}_{12}) + \boldsymbol{n}_3 (\boldsymbol{\phi}_r, \boldsymbol{\phi}_r, \boldsymbol{\phi}_r)], \tag{6.33c}$$

$$\mathbf{q}_{21} := - i\omega_r A_r \mathcal{D}\boldsymbol{\phi}_r + \frac{1}{2}\alpha \mathbf{f}_0 e^{i\sigma_r t_2} \tag{6.33d}$$

are resonant and non-resonant terms (by assuming $3\omega_r \neq \omega_k \ \forall k$). However, the non-resonant ones, $\mathbf{p}_{22}, \mathbf{p}_{23}$, are not actually useful, if the analysis is truncated at this order. Indeed, compatibility requires that:

$$\int_\Omega \boldsymbol{\phi}_r^T \mathbf{p}_{21} \, d\Omega + \int_\Gamma \boldsymbol{\phi}_r^T \mathbf{q}_{21} \, d\Gamma = 0, \tag{6.34}$$

from which an ordinary differential equation for $A_r(t_2)$ follows. Since $\dot{A}_r = (\cancel{\partial_0} + \epsilon \cancel{\partial_1} + \epsilon^2 \partial_2) A_r$, by coming back to the unrescaled quantities, it reads:

$$\dot{A}_r = -c_1 A_r + i c_3 A_r^2 \bar{A}_r + i c_0 \alpha e^{i\sigma_r t}, \tag{6.35}$$

where the c_k are real constants, defined as follows:

$$c_1 := \frac{\int_\Omega \boldsymbol{\phi}_r^T \mathcal{C}\boldsymbol{\phi}_r \, \mathrm{d}\Omega + \int_\Gamma \boldsymbol{\phi}_r^T \mathcal{D}\boldsymbol{\phi}_r \, \mathrm{d}\Gamma}{2\int_\Omega \boldsymbol{\phi}_r^T \mathcal{M}\boldsymbol{\phi}_r \, \mathrm{d}\Omega}, \quad c_0 := -\frac{\int_\Omega \boldsymbol{\phi}_r^T \mathbf{b}_0 \, \mathrm{d}\Omega + \int_\Gamma \boldsymbol{\phi}_r^T \mathbf{f}_0 \, \mathrm{d}\Gamma}{4\omega_r \int_\Omega \boldsymbol{\phi}_r^T \mathcal{M}\boldsymbol{\phi}_r \, \mathrm{d}\Omega},$$

$$c_3 := \frac{1}{2\omega_r \int_\Omega \boldsymbol{\phi}_r^T \mathcal{M}\boldsymbol{\phi}_r \, \mathrm{d}\Omega} \int_\Omega \boldsymbol{\phi}_r^T \Big[2\boldsymbol{n}_2(\boldsymbol{\phi}_r, \boldsymbol{\chi}_{10}) + 2\boldsymbol{n}_2(\boldsymbol{\phi}_r, \boldsymbol{\chi}_{12}) \tag{6.36}$$

$$+ 2\boldsymbol{n}_2(\boldsymbol{\phi}_r, \bar{\boldsymbol{\chi}}_{10}) + 3\boldsymbol{n}_3(\boldsymbol{\phi}_r, \boldsymbol{\phi}_r, \boldsymbol{\phi}_r) \Big] \mathrm{d}\Omega.$$

This is the bifurcation equation ruling the evolution of amplitude and phase of the leading part of the response, Eq. 6.29.[14] Worked examples of structures involved in external resonance will be presented in the Problems 7.5, 7.6 and 7.7.

6.4.2 Sub- and Super-Harmonic Resonances

The sub-harmonic resonance (occurring when $\Omega \simeq 3\omega_r$) and the super-harmonic resonance (taking place when $3\Omega \simeq \omega_r$) are now studied. The investigation is limited to symmetric systems, for which $\boldsymbol{n}_2(\mathbf{u}, \mathbf{u}) \equiv \mathbf{0}$.

Multiple Scale Perturbation Equations

The following rescaling is introduced: $\mathbf{u} \to \epsilon^{1/2}\hat{\mathbf{u}}$, $(\mathcal{C}, \mathcal{D}) \to \epsilon\left(\hat{\mathcal{C}}, \hat{\mathcal{D}}\right)$, while, differently from the previous case, the load is considered to be large, of the same order of the response, i.e., $\alpha = \epsilon^{1/2}\hat{\alpha}$ (*hard excitation*). The rescaled Eqs. 6.20 assume the form:

$$\mathcal{M}\ddot{\mathbf{u}} + \mathcal{K}\mathbf{u} = -\epsilon\left[\mathcal{C}\dot{\mathbf{u}} + \boldsymbol{n}_3(\mathbf{u}, \mathbf{u}, \mathbf{u})\right] + \left(\frac{1}{2}\alpha\mathbf{b}_0 e^{i\Omega t} + \text{c.c.}\right), \tag{6.37a}$$

$$\mathcal{B}\mathbf{u} = -\epsilon\mathcal{D}\dot{\mathbf{u}} + \left(\frac{1}{2}\alpha\mathbf{f}_0 e^{i\Omega t} + \text{c.c.}\right), \tag{6.37b}$$

The displacement is expanded as $\mathbf{u} = \mathbf{u}_0 + \epsilon\mathbf{u}_1 + \epsilon^2\mathbf{u}_2 + \cdots$, and independent time scales $t_0 := t$, $t_1 := \epsilon t, \ldots$ are introduced, so that $\dot{\mathbf{u}} = (\partial_0 + \epsilon\partial_1 + \cdots)\,\mathbf{u}$, $\ddot{\mathbf{u}} = (\partial_0 + \epsilon\partial_1 + \cdots)^2\,\mathbf{u}$. The following perturbation equations are derived:

Order ϵ^0:

$$\mathcal{M}\partial_0^2\mathbf{u}_0 + \mathcal{K}\mathbf{u}_0 = \frac{1}{2}\alpha\mathbf{b}_0 e^{i\Omega t_0} + \text{c.c.}, \tag{6.38a}$$

$$\mathcal{B}\mathbf{u}_0 = \frac{1}{2}\alpha\mathbf{f}_0 e^{i\Omega t_0} + \text{c.c.}. \tag{6.38b}$$

[14] The motion, therefore, takes place on a family of two dimensional manifolds, parametrized by α and σ_r.

Order ϵ^1:

$$\mathcal{M}\partial_0^2\mathbf{u}_1 + \mathcal{K}\mathbf{u}_1 = -2\mathcal{M}\partial_0\partial_1\mathbf{u}_0 - \mathcal{C}\partial_0\mathbf{u}_0 - \boldsymbol{n}_3(\mathbf{u}_0, \mathbf{u}_0, \mathbf{u}_0), \tag{6.39a}$$

$$\mathcal{B}\mathbf{u}_1 = -\mathcal{D}\partial_0\mathbf{u}_0. \tag{6.39b}$$

Solution

The generating solution is:

$$\mathbf{u}_0 = A_r(t_1, \ldots)\,\boldsymbol{\phi}_r(\mathbf{s})\,e^{i\omega_r t_0} + \frac{1}{2}\alpha\boldsymbol{\chi}_0(\mathbf{s})\,e^{i\Omega t_0} + \text{c.c.}, \tag{6.40}$$

where $\boldsymbol{\chi}_0(\mathbf{s})$ is uniquely determined by a non-homogeneous boundary value problem, governed by the non-singular operator $\mathcal{K} - \Omega^2\mathcal{M}$. When Eq. 6.38 is substituted in Eq. 6.39, this latter becomes:

$$\mathcal{M}\partial_0^2\mathbf{u}_1 + \mathcal{K}\mathbf{u}_1 = \mathbf{p}_{11}(\mathbf{s})\,e^{i\omega_r t_0} + \mathbf{p}_{12}(\mathbf{s})\,e^{3i\Omega t_0} + \mathbf{p}_{13}(\mathbf{s})\,e^{i(\Omega-2\omega_r)t_0} + \text{c.c.} + \text{N.R.T.}, \tag{6.41a}$$

$$\mathcal{B}\mathbf{u}_1 = \mathbf{q}_{11}(\mathbf{s})\,e^{i\omega_r t_0} + \text{c.c.} + \text{N.R.T.}, \tag{6.41b}$$

where:

$$\mathbf{p}_{11} := -2i\omega_r\partial_1 A_r\mathcal{M}\boldsymbol{\phi}_r - i\omega_r A_r\mathcal{C}\boldsymbol{\phi}_r - 3A_r^2\bar{A}_r\boldsymbol{n}_3\left(\boldsymbol{\phi}_r, \boldsymbol{\phi}_r, \boldsymbol{\phi}_r\right) - \frac{3}{2}\alpha^2 A_r\boldsymbol{n}_3\left(\boldsymbol{\phi}_r, \boldsymbol{\chi}_0, \boldsymbol{\chi}_0\right), \tag{6.42a}$$

$$\mathbf{p}_{12} := -\frac{1}{8}\alpha^3\boldsymbol{n}_3\left(\boldsymbol{\chi}_0, \boldsymbol{\chi}_0, \boldsymbol{\chi}_0\right), \tag{6.42b}$$

$$\mathbf{p}_{13} := -\frac{3}{2}\alpha\bar{A}_r^2\boldsymbol{n}_3\left(\boldsymbol{\phi}_r, \boldsymbol{\phi}_r, \boldsymbol{\chi}_0\right), \tag{6.42c}$$

$$\mathbf{q}_{11} := -i\omega_r A_r\mathcal{D}\boldsymbol{\phi}_r \tag{6.42d}$$

and the nonresonant terms collect the frequencies $\Omega, 3\omega_r, \Omega + 2\omega_r, 2\Omega \pm \omega_r$. The two resonances are now separately analyzed.

Sub-harmonic Resonance

In the sub-harmonic case, $\Omega = 3\omega_r + \epsilon\hat{\sigma}_r$ is introduced in the $\mathbf{p}_{13}$-term of Eq. 6.41 and the following compatibility condition enforced:

$$\int_{\Omega} \boldsymbol{\phi}_r^T \left(\mathbf{p}_{11} + \mathbf{p}_{13} e^{i\sigma_r t_1} \right) \mathrm{d}\Omega + \int_{\Gamma} \boldsymbol{\phi}_r^T \mathbf{q}_{11} \mathrm{d}\Gamma = 0, \tag{6.43}$$

from which the bifurcation equation is drawn:

$$\dot{A}_r = -c_1 A_r + i c_3 A_r^2 \bar{A}_r + i c_{21} \alpha^2 A_r + i c_{12} \alpha \bar{A}_r^2 e^{i\sigma_r t}, \tag{6.44}$$

where the following definitions hold:

$$c_1 := \frac{\int_\Omega \boldsymbol{\phi}_r^T \boldsymbol{\mathcal{C}} \boldsymbol{\phi}_r \mathrm{d}\Omega + \int_\Gamma \boldsymbol{\phi}_r^T \boldsymbol{\mathcal{D}} \boldsymbol{\phi}_r \mathrm{d}\Gamma}{2 \int_\Omega \boldsymbol{\phi}_r^T \boldsymbol{\mathcal{M}} \boldsymbol{\phi}_r \mathrm{d}\Omega}, \quad c_3 := \frac{3 \int_\Omega \boldsymbol{\phi}_r^T \boldsymbol{n}_3(\boldsymbol{\phi}_r, \boldsymbol{\phi}_r, \boldsymbol{\phi}_r) \mathrm{d}\Omega}{2\omega_r \int_\Omega \boldsymbol{\phi}_r^T \boldsymbol{\mathcal{M}} \boldsymbol{\phi}_r \mathrm{d}\Omega}, \tag{6.45a}$$

$$c_{21} := \frac{3 \int_\Omega \boldsymbol{\phi}_r^T \boldsymbol{n}_3(\boldsymbol{\phi}_r, \boldsymbol{\chi}_0, \boldsymbol{\chi}_0) \mathrm{d}\Omega}{4\omega_r \int_\Omega \boldsymbol{\phi}_r^T \boldsymbol{\mathcal{M}} \boldsymbol{\phi}_r \mathrm{d}\Omega}, \quad c_{12} := \frac{3 \int_\Omega \boldsymbol{\phi}_r^T \boldsymbol{n}_3(\boldsymbol{\phi}_r, \boldsymbol{\phi}_r, \boldsymbol{\chi}_0) \mathrm{d}\Omega}{4\omega_r \int_\Omega \boldsymbol{\phi}_r^T \boldsymbol{\mathcal{M}} \boldsymbol{\phi}_r \mathrm{d}\Omega}. \tag{6.45b}$$

Super-Harmonic Resonance

In the super-harmonic case, $3\Omega = \omega_r + \epsilon \hat{\sigma}_r$ is introduced in the $\mathbf{p}_{12}$-term of Eq. 6.41 and the compatibility condition imposed:

$$\int_{\Omega} \boldsymbol{\phi}_r^T \left(\mathbf{p}_{11} + \mathbf{p}_{12} e^{i\sigma_r t_1} \right) \mathrm{d}\Omega + \int_{\Gamma} \boldsymbol{\phi}_r^T \mathbf{q}_{11} \mathrm{d}\Gamma = 0. \tag{6.46}$$

This provides the following bifurcation equation:

$$\dot{A}_r = -c_1 A_r + i c_3 A_r^2 \bar{A}_r + i c_{21} \alpha^2 A_r + i c_{30} \alpha^3 e^{i\sigma_r t}, \tag{6.47}$$

where c_1, c_3 and c_{21} are still given by Eqs. 6.45, while:

$$c_{30} := \frac{\int_\Omega \boldsymbol{\phi}_r^T \boldsymbol{n}_3(\boldsymbol{\chi}_0, \boldsymbol{\chi}_0, \boldsymbol{\chi}_0) \mathrm{d}\Omega}{16\omega_r \int_\Omega \boldsymbol{\phi}_r^T \boldsymbol{\mathcal{M}} \boldsymbol{\phi}_r \mathrm{d}\Omega}. \tag{6.48}$$

6.5 Principal Parametric Excitation

The principal parametric excitation has been discussed in the Sect. 3.5 by referring to the sample system, as an illustrative example; now, the analysis is extended to general structures. The problem is tackled by the MSM [1, 6–8, 11, 12, 14]; a different approach, based on the Harmonic Balance Method, is illustrated in [3], by dealing with the Bolotin beam.

When a system possesses periodically time-variant coefficients, as it happens, e.g., in case of pulsating forces which induce geometric effects, the system is said to be *parametrically excited*. Depending on the ratio between the parametric frequency and one of the natural frequencies of the system, as well as on the amplitude of the excitation, the trivial equilibrium position, admitted to exist, can lose stability. The scope of the analysis is: (a) to find the instability region of the trivial response in the parameter plane, and (b) to detect the existence of stable limit cycles, able to confine the response.

The equations of motion governing parametrically excited damped elastic systems are of type of Eq. 5.3, in which, however, the geometric forces are periodic in time, i.e.:

$$\mathcal{M}\ddot{\mathbf{u}} + \mathcal{C}_s\dot{\mathbf{u}} + \left[\mathcal{K}_e + \mu\cos(\Omega t)\,\mathcal{K}_g\right]\mathbf{u} + \boldsymbol{n}(\mathbf{u};\mu) = \mathbf{0}, \qquad \text{in } \Omega, \tag{6.49a}$$

$$\mathcal{D}_s\dot{\mathbf{u}} + \left(\mathcal{B}_s + \mu\cos(\Omega t)\,\mathcal{B}_g\right)\mathbf{u} = \mathbf{0}, \qquad \text{on } \Gamma. \tag{6.49b}$$

Here, μ is the amplitude of the parametric excitation and Ω its frequency, which are left as free parameters. In what follows, for the sake of simplicity, it will be assumed that nonlinearities are only cubic (i.e., the system behaves symmetrically), i.e., $\boldsymbol{n}(\mathbf{u};\mu) = \boldsymbol{n}_3(\mathbf{u},\mathbf{u},\mathbf{u};\mu) + \cdots$. Nonlinearities, in principle, are also affected by the parametric excitation but, as it will appear clear soon, these do not enter into a low-order solution.

Multiple Scale Perturbation Equations

The MSM is used to tackle Eqs. 6.49. First, the rescaling $\mathbf{u} \to \epsilon^{1/2}\mathbf{u}$ is used, to denote the smallness of the response,[15] together with $\mu \to \epsilon\hat{\mu}$, to consider parametric excitation of small amplitude. Moreover, damping is taken small, via $(\mathcal{C}_s, \mathcal{D}_s) \to \epsilon\left(\hat{\mathcal{C}}_s, \hat{\mathcal{D}}_s\right)$. The rescaled equations (with hat omitted), then read:

$$\mathcal{M}\ddot{\mathbf{u}} + \mathcal{K}_e\mathbf{u} = -\epsilon\left[\mathcal{C}_s\dot{\mathbf{u}} + \mu\cos(\Omega t)\,\mathcal{K}_g\mathbf{u} + \boldsymbol{n}_3(\mathbf{u},\mathbf{u},\mathbf{u};0)\right], \tag{6.50a}$$

$$\mathcal{B}_s\mathbf{u} = -\epsilon\left[\mathcal{D}_s\dot{\mathbf{u}} + \mu\cos(\Omega t)\,\mathcal{B}_g\mathbf{u}\right]. \tag{6.50b}$$

By expanding the displacements as $\mathbf{u} = \mathbf{u}_0 + \epsilon\mathbf{u}_1 + \cdots$, and introducing the time-scales $t_0 := t,\ t_1 := \epsilon t, \ldots$, for which $\dot{\mathbf{u}} = (\partial_0 + \epsilon\partial_1 + \cdots)\,\mathbf{u}$, $\ddot{\mathbf{u}} = (\partial_0 + \epsilon\partial_1 + \cdots)^2\,\mathbf{u}$, the following perturbation equations are derived:

Order ϵ^0:

$$\mathcal{M}\partial_0^2\mathbf{u}_0 + \mathcal{K}_e\mathbf{u}_0 = \mathbf{0}, \tag{6.51a}$$

$$\mathcal{B}_s\mathbf{u}_0 = \mathbf{0}. \tag{6.51b}$$

[15] This rescaling is useful for symmetric systems, to make $|\mathbf{u}|^3 = O(\epsilon\,|\mathbf{u}|)$.

Order ϵ^1:

$$\mathcal{M}\partial_0^2\mathbf{u}_1 + \mathcal{K}_e\mathbf{u}_1 = -\Big[2\mathcal{M}\partial_0\partial_1\mathbf{u}_0 + \mathcal{C}_s\partial_0\mathbf{u}_0 + \frac{1}{2}\mu\left(e^{i\Omega t_0} + \text{c.c.}\right)\mathcal{K}_g\mathbf{u} \\ + \boldsymbol{n}_3\left(\mathbf{u}_0, \mathbf{u}_0, \mathbf{u}_0; 0\right)\Big], \tag{6.52a}$$

$$\mathcal{B}_s\mathbf{u}_1 = -\left[\mathcal{D}_s\partial_0\mathbf{u}_0 + \frac{1}{2}\mu\left(e^{i\Omega t_0} + \text{c.c.}\right)\mathcal{B}_g\mathbf{u}_0\right], \tag{6.52b}$$

where the complex form has been introduced for $\cos(\Omega t_0)$.

Solution

A monomodal solution to Eq. 6.51 is taken, namely:

$$\mathbf{u}_0 = A_r\left(t_1, \ldots\right)\boldsymbol{\phi}_r\left(\mathbf{s}\right)e^{i\omega_r t_0} + \text{c.c.}, \tag{6.53}$$

where ω_r is a natural frequency in *principal parametric resonance* with Ω (to be determined, yet), and $\boldsymbol{\phi}_r$ is the (real) associated natural mode,[16] all other frequencies being damped and supposed not to be in internal resonance. When Eq. 6.53 is substituted in the ϵ order perturbation Eqs. 6.52, this latter become:

$$\mathcal{M}\partial_0^2\mathbf{u}_1 + \mathcal{K}_e\mathbf{u}_1 = -\Big[2i\omega_r\partial_1 A_r\mathcal{M}\boldsymbol{\phi}_r + i\omega_r A_r\mathcal{C}_s\boldsymbol{\phi}_r \\ + 3A_r^2\bar{A}_r\boldsymbol{n}_3\left(\boldsymbol{\phi}_r, \boldsymbol{\phi}_r, \boldsymbol{\phi}_r; 0\right)\Big]e^{i\omega_r t_0} - \frac{1}{2}\mu\Big[A_r e^{i(\Omega+\omega_r)t_0} \\ + \bar{A}_r e^{i(\Omega-\omega_r)t_0}\Big]\mathcal{K}_g\boldsymbol{\phi}_r + \text{N.R.T.} + \text{c.c.}, \tag{6.54a}$$

$$\mathcal{B}_s\mathbf{u}_1 = -i\omega_r A_r\mathcal{D}_s\boldsymbol{\phi}_r e^{i\omega_r t_0} \\ - \frac{1}{2}\mu\left[A_r e^{i(\Omega+\omega_r)t_0} + \bar{A}_r e^{i(\Omega-\omega_r)t_0}\right]\mathcal{B}_g\boldsymbol{\phi}_r + \text{c.c.}, \tag{6.54b}$$

where the non-resonant terms (N.R.T.) generated by the cubic nonlinearities have been ignored.

Concerning the parametric excitation terms, the relationship between Ω and ω_r needs to be investigated. The frequency $\Omega + \omega_r$ is certainly non-resonant, since different from ω_r; the frequency $\Omega - \omega_r$, instead, is resonant if $\Omega \simeq 2\omega_r$ and non-resonant if $\Omega \neq 2\omega_r$. Therefore, the parametric frequency $\Omega \simeq 2\omega_r$ only excites, at this order, the rth natural

[16] The adjective 'principal' refers to the fact that the resonance is the first one encountered in the perturbation scheme, and therefore more important, since of lowest order. 'Secondary' resonances would appear to higher orders, not investigated here (see [11]).

mode. To explore the neighborhood of the perfect principal parametric resonance $\Omega = 2\omega_r$, a small detuning is introduced, by letting $\Omega = 2\omega_r + \epsilon\sigma_r$; accordingly, Eqs. 6.54 become:

$$\mathcal{M}\partial_0^2\mathbf{u}_1 + \mathcal{K}_e\mathbf{u}_1 = \mathbf{p}_{11}e^{i\omega_r t_0} + \text{N.R.T.} + \text{c.c.}, \tag{6.55a}$$

$$\mathcal{B}_s\mathbf{u}_1 = \mathbf{q}_{11}e^{i\omega_r t_0} + \text{N.R.T.} + \text{c.c.}, \tag{6.55b}$$

where:

$$\begin{aligned}\mathbf{p}_{11} :=& -2i\omega_r\partial_1 A_r\mathcal{M}\boldsymbol{\phi}_r - i\omega_r A_r\mathcal{C}_s\boldsymbol{\phi}_r - 3A_r^2\bar{A}_r\boldsymbol{n}_3\left(\boldsymbol{\phi}_r, \boldsymbol{\phi}_r, \boldsymbol{\phi}_r; 0\right) \\ &- \frac{1}{2}\mu\bar{A}_r e^{i\sigma_r t_1}\mathcal{K}_g\boldsymbol{\phi}_r,\end{aligned} \tag{6.56a}$$

$$\mathbf{q}_{11} := -i\omega_r A_r\mathcal{D}_s\boldsymbol{\phi}_r - \frac{1}{2}\mu\bar{A}_r e^{i\sigma_r t_1}\mathcal{B}_g\boldsymbol{\phi}_r. \tag{6.56b}$$

The compatibility condition for Eq. 6.55 requires that:

$$\int_\Omega \boldsymbol{\phi}_r^T\mathbf{p}_{11}\,\mathrm{d}\Omega + \int_\Gamma \boldsymbol{\phi}_r^T\mathbf{q}_{11}\,\mathrm{d}\Gamma = 0, \tag{6.57}$$

from which, by coming back to the true time, via $\dot{A}_r = \epsilon\partial_1 A_r$, and to the unrescaled quantities, the bifurcation equation is drawn:

$$\dot{A}_r = -c_1 A_r + ic_3 A_r^2\bar{A}_r + ic_0\mu\bar{A}_r e^{i\sigma_r t}, \tag{6.58}$$

where the real constants c_k are as follows:

$$\begin{aligned} c_1 &:= \frac{\int_\Omega \boldsymbol{\phi}_r^T\mathcal{C}_s\boldsymbol{\phi}_r\mathrm{d}\Omega + \int_\Gamma \boldsymbol{\phi}_r^T\mathcal{D}_s\boldsymbol{\phi}_r\mathrm{d}\Gamma}{2\int_\Omega \boldsymbol{\phi}_r^T\mathcal{M}\boldsymbol{\phi}_r\mathrm{d}\Omega}, \quad c_3 := \frac{3\int_\Omega \boldsymbol{\phi}_r^T\boldsymbol{n}_3\left(\boldsymbol{\phi}_r, \boldsymbol{\phi}_r, \boldsymbol{\phi}_r; 0\right)\mathrm{d}\Omega}{2\omega_r\int_\Omega \boldsymbol{\phi}_r^T\mathcal{M}\boldsymbol{\phi}_r\mathrm{d}\Omega}, \\ c_0 &:= \frac{\int_\Omega \boldsymbol{\phi}_r^T\mathcal{K}_g\boldsymbol{\phi}_r\mathrm{d}\Omega + \int_\Gamma \boldsymbol{\phi}_r^T\mathcal{B}_g\boldsymbol{\phi}_r\mathrm{d}\Gamma}{4\omega_r\int_\Omega \boldsymbol{\phi}_r^T\mathcal{M}\boldsymbol{\phi}_r\mathrm{d}\Omega}. \end{aligned} \tag{6.59}$$

Equation 6.58 governs the amplitude and phase modulation of the response in Eq. 6.53.[17] An example of parametric excitation analysis is presented in the Problem 7.9.

[17] The motion, therefore, develops on a family of two dimensional manifolds, parametrized by μ and σ_r.

6.6 Dynamic Bifurcation

Dynamic bifurcations, already discussed in the Sect. 4.2 with reference to the sample system, are now reconsidered and studied for a general system. The solution is pursued by the MSM [1, 6–8, 11, 12, 14].

When an elastic system is subject to nonconservative forces, its stiffness operator $\mathcal{K}$ (and/or the associate boundary conditions $\mathcal{B}$), which includes the effect of the external forces, is non-self-adjoint. When the external force, taken proportional to a multiplier μ, exceeds a critical value μ_c, the trivial fundamental path, admitted to exist, loses stability via a *dynamic* (or Hopf) *bifurcation*, at which a family of *limit cycles*, parametrized by $\mu - \mu_c$, arises. A paradigmatic example is offered by the Beck beam [2, 9]. Task of the analysis is to find the bifurcation point, to build-up the family of limit cycles and to analyze their stability.

The equations of motion of the system are of type in Eq. 5.3, i.e.:

$$\mathcal{M}\ddot{\mathbf{u}} + \mathcal{C}_s\dot{\mathbf{u}} + \left(\mathcal{K}_e + \mu\mathcal{K}_g\right)\mathbf{u} + \boldsymbol{n}\left(\mathbf{u};\mu\right) = \mathbf{0}, \qquad \text{in } \Omega, \tag{6.60a}$$

$$\mathcal{D}_s\dot{\mathbf{u}} + \left(\mathcal{B}_s + \mu\mathcal{B}_g\right)\mathbf{u} = \mathbf{0}, \qquad \text{on } \Gamma. \tag{6.60b}$$

where $\boldsymbol{n}\left(\mathbf{u};\mu\right) = \boldsymbol{n}_2\left(\mathbf{u},\mathbf{u};\mu\right) + \boldsymbol{n}_3\left(\mathbf{u},\mathbf{u},\mathbf{u};\mu\right) + \cdots$. Here, the geometric effects on damping have been ignored,[18] and known external forces assumed absent.[19]

Multiple Scale Perturbation Equations

To apply a perturbation method to Eqs. 6.60, the displacement is rescaled as $\mathbf{u} \to \epsilon\hat{\mathbf{u}}$; moreover, the load parameter is written as $\mu = \mu_0 + \epsilon^2\tilde{\mu}$ in order to explore the neighborhood of the unknown bifurcation value μ_0.[20] Differently from what done in the Sect. 6.4, the damping is not rescaled here, to avoid some degenerateness (see Note 5 in Chap. 4). Hence, the rescaled equations (with hat omitted), read[21]:

$$\begin{aligned}\mathcal{M}\ddot{\mathbf{u}} + \mathcal{C}_s\dot{\mathbf{u}} + \left(\mathcal{K}_e + \mu_0\mathcal{K}_g\right)\mathbf{u} = &-\epsilon\boldsymbol{n}_2\left(\mathbf{u},\mathbf{u};\mu_0\right) \\ &-\epsilon^2\left[\tilde{\mu}\mathcal{K}_g\mathbf{u} + \boldsymbol{n}_3\left(\mathbf{u},\mathbf{u},\mathbf{u};\mu_0\right)\right],\end{aligned} \tag{6.61a}$$

$$\mathcal{D}_s\dot{\mathbf{u}} + \left(\mathcal{B}_s + \mu_0\mathcal{B}_g\right)\mathbf{u} = -\epsilon^2\tilde{\mu}\mathcal{B}_g\mathbf{u}. \tag{6.61b}$$

---

[18] They are typical of the aeroelastic forces [10], not addressed in this book.

[19] They have no *qualitative* influence on dynamic bifurcations [10].

[20] This operation is called *splitting* of the parameter, which is something different from expansion (which contains an infinite number of terms).

[21] Terms as $\boldsymbol{n}_{2,\mu}\left(\mathbf{u},\mathbf{u};\mu_0\right)$, mentioned in the Note 4 of this chapter by dealing with buckling, are here of higher order, since the neighborhood of the bifurcation point explored is now smaller, of order ϵ^2.

The MSM is applied, as described by Eqs. 6.24, 6.25, but omitting the t_1 time-scale, which does not contribute to the solution, as already seen in Sect. 6.4; accordingly $\dot{\mathbf{u}} = \left(\partial_0 + \epsilon^2\partial_2 + \cdots\right)\mathbf{u}$, $\ddot{\mathbf{u}} = \left(\partial_0 + \epsilon^2\partial_2 + \cdots\right)^2\mathbf{u}$. Moreover, $\mathbf{u}$ is expanded as $\mathbf{u} = \mathbf{u}_0 + \epsilon\mathbf{u}_1 + \epsilon^2\mathbf{u}_2 + \cdots$. The following perturbation equations are thus derived:

Order ϵ^0:

$$\mathcal{M}\partial_0^2\mathbf{u}_0 + \mathcal{C}_s\partial_0\mathbf{u}_0 + \left(\mathcal{K}_e + \mu_0\mathcal{K}_g\right)\mathbf{u}_0 = \mathbf{0}, \tag{6.62a}$$
$$\mathcal{D}_s\partial_0\mathbf{u}_0 + \left(\mathcal{B}_s + \mu_0\mathcal{B}_g\right)\mathbf{u}_0 = \mathbf{0}. \tag{6.62b}$$

Order ϵ^1:

$$\mathcal{M}\partial_0^2\mathbf{u}_1 + \mathcal{C}_s\partial_0\mathbf{u}_1 + \left(\mathcal{K}_e + \mu_0\mathcal{K}_g\right)\mathbf{u}_1 = -\boldsymbol{n}_2\left(\mathbf{u}_0, \mathbf{u}_0; \mu_0\right), \tag{6.63a}$$
$$\mathcal{D}_s\partial_0\mathbf{u}_1 + \left(\mathcal{B}_s + \mu_0\mathcal{B}_g\right)\mathbf{u}_1 = \mathbf{0}. \tag{6.63b}$$

Order ϵ^2:

$$\begin{aligned}\mathcal{M}\partial_0^2\mathbf{u}_2 + \mathcal{C}_s\partial_0\mathbf{u}_2 + \left(\mathcal{K}_e + \mu_0\mathcal{K}_g\right)\mathbf{u}_2 &= -2\mathcal{M}\partial_0\partial_2\mathbf{u}_0 - \mathcal{C}_s\partial_2\mathbf{u}_0 \\ &\quad - \tilde{\mu}\mathcal{K}_g\mathbf{u}_0 - 2\boldsymbol{n}_2\left(\mathbf{u}_0, \mathbf{u}_1; \mu_0\right) \\ &\quad - \boldsymbol{n}_3\left(\mathbf{u}_0, \mathbf{u}_0, \mathbf{u}_0; \mu_0\right),\end{aligned} \tag{6.64a}$$
$$\mathcal{D}_s\partial_0\mathbf{u}_2 + \left(\mathcal{B}_s + \mu_0\mathcal{B}_g\right)\mathbf{u}_2 = -\mathcal{D}_s\partial_2\mathbf{u}_0 - \tilde{\mu}\mathcal{B}_g\mathbf{u}_0. \tag{6.64b}$$

Generating Solution

The generating Eqs. 6.62 are addressed. By letting $\mathbf{u} = \boldsymbol{\phi}_k e^{\lambda_k t_0}$, the space-boundary value problem follows:

$$\left(\mathcal{K}_e + \mu_0\mathcal{K}_g + \lambda_k\mathcal{C}_s + \lambda_k^2\mathcal{M}\right)\boldsymbol{\phi}_k = \mathbf{0}, \tag{6.65a}$$
$$\left(\mathcal{B}_s + \mu_0\mathcal{B}_g + \lambda_k\mathcal{D}_s\right)\boldsymbol{\phi}_k = \mathbf{0}. \tag{6.65b}$$

Since the operator acting on $\boldsymbol{\phi}_k$ is non-self-adjoint, the eigenpairs $\left(\lambda_k, \boldsymbol{\phi}_k\right)$ are generally complex, together with their conjugate $\left(\bar{\lambda}_k, \bar{\boldsymbol{\phi}}_k\right)$. When μ_0 is small, or even zero, being $\mathcal{K}_e$ definite positive and $\mathcal{C}_s$ dissipative by hypothesis (together to their boundary counterparts), it is $\mathrm{Re}\left(\lambda_k\right) < 0\ \forall k$, so that the equilibrium is asymptotically stable. However, when μ_0 reaches a critical value μ_c, it is likely to occur that a pair of complex conjugate eigenvalues becomes purely imaginary and, for a further infinitesimal increment of the load, they cross the imaginary axis, this entailing equilibrium instability (and Hopf bifurcation). By denoting one of the two critical eigenvalues by $\lambda_c = i\omega_c$, and the associated complex critical mode by $\boldsymbol{\phi}_c$, the following equations hold for the critical quantities:

$$\left(\mathcal{K}_e + \mu_c \mathcal{K}_g + i\omega_c \mathcal{C}_s - \omega_c^2 \mathcal{M}\right) \boldsymbol{\phi}_c = \mathbf{0}, \tag{6.66a}$$

$$\left(\mathcal{B}_s + \mu_c \mathcal{B}_g + i\omega_c \mathcal{D}_s\right) \boldsymbol{\phi}_c = \mathbf{0}. \tag{6.66b}$$

The solution to Eqs. 6.62 therefore reads[22]:

$$\mathbf{u}_0 = A\left(t_2, \ldots\right) \boldsymbol{\phi}_c\left(\mathbf{s}\right) e^{i\omega_c t_0} + \text{c.c.}. \tag{6.67}$$

Solution to Higher Order Perturbation Equations

By substituting Eq. 6.67 in the ϵ order perturbation Eqs. 6.63, the latter read:

$$\mathcal{M}\partial_0^2 \mathbf{u}_1 + \mathcal{C}_s \partial_0 \mathbf{u}_1 + \left(\mathcal{K}_e + \mu_c \mathcal{K}_g\right) \mathbf{u}_1 = -A^2 e^{2i\omega_c t_0} \boldsymbol{n}_2\left(\boldsymbol{\phi}_c, \boldsymbol{\phi}_c; \mu_c\right) - A\bar{A}\boldsymbol{n}_2\left(\boldsymbol{\phi}_c, \bar{\boldsymbol{\phi}}_c; \mu_c\right) + \text{c.c.}, \tag{6.68a}$$

$$\mathcal{D}_s \partial_0 \mathbf{u}_1 + \left(\mathcal{B}_s + \mu_c \mathcal{B}_g\right) \mathbf{u}_1 = \mathbf{0}. \tag{6.68b}$$

The relevant solution, by excluding any internal resonance, is:

$$\mathbf{u}_1 = A^2 \boldsymbol{\chi}_{12}(\mathbf{s}) e^{2i\omega_c t_0} + A\bar{A}\boldsymbol{\chi}_{10}(\mathbf{s}) + \text{c.c.}, \tag{6.69}$$

where, due to the linearity, the $\boldsymbol{\chi}\left(\mathbf{s}\right)$ functions are unique solutions to boundary value problems governed by non-singular operators, namely $\mathcal{L}_{12} := \left(\mathcal{K}_e + \mu_c \mathcal{K}_g\right) - 4\omega_c^2 \mathcal{M} + i\omega_c \mathcal{C}_s$, and $\mathcal{L}_{10} := \mathcal{K}_e + \mu_c \mathcal{K}_g$, respectively (and similar expressions at the boundary).

As Eqs. 6.67 and 6.69 are substituted in the ϵ^2 order perturbation Eqs. 6.64, these assume the form:

$$\mathcal{M}\partial_0^2 \mathbf{u}_2 + \mathcal{C}_s \partial_0 \mathbf{u}_2 + \left(\mathcal{K}_e + \mu_c \mathcal{K}_g\right) \mathbf{u}_2 = \mathbf{p}_{21} e^{i\omega_c t_0} + \text{N.R.T.} + \text{c.c.}, \tag{6.70a}$$

$$\mathcal{D}_s \partial_0 \mathbf{u}_2 + \left(\mathcal{B}_s + \mu_c \mathcal{B}_g\right) \mathbf{u}_2 = \mathbf{q}_{21} e^{i\omega_c t_0} + \text{N.R.T.} + \text{c.c.}, \tag{6.70b}$$

[22] Equation 6.67 is formally identical to Eq. 6.29, relevant to the external resonance of a conservative system, but now the mode $\boldsymbol{\phi}_c$ is complex, entailing:

$$\mathbf{u}_0 = a\left(t_2\right)\left\{\text{Re}\left(\boldsymbol{\phi}_c\right)\cos\left[\omega_c t_0 + \varphi\left(t_2\right)\right] - \text{Im}\left(\boldsymbol{\phi}_c\right)\sin\left[\omega_c t_0 + \varphi\left(t_2\right)\right]\right\},$$

as $A(t_2) = \frac{1}{2}a(t_2)e^{i\varphi(t_2)}$.

where:

$$\begin{aligned}\mathbf{p}_{21} :=& -\tilde{\mu}A\mathcal{K}_g\boldsymbol{\phi}_c - \partial_2 A\mathcal{C}_s\boldsymbol{\phi}_c - 2i\omega_c\partial_2 A\mathcal{M}\boldsymbol{\phi}_c \\ & - 3A^2\bar{A}\boldsymbol{n}_3\left(\boldsymbol{\phi}_c, \boldsymbol{\phi}_c, \bar{\boldsymbol{\phi}}_c, \mu_c\right) - 2A^2\bar{A}\boldsymbol{n}_2\left(\boldsymbol{\phi}_c, \boldsymbol{\chi}_{10}, \mu_c\right) \\ & - 2A^2\bar{A}\boldsymbol{n}_2\left(\boldsymbol{\phi}_c, \bar{\boldsymbol{\chi}}_{10}, \mu_c\right) - 2A^2\bar{A}\boldsymbol{n}_2\left(\bar{\boldsymbol{\phi}}_c, \boldsymbol{\chi}_{12}, \mu_c\right),\end{aligned} \tag{6.71a}$$

$$\mathbf{q}_{21} := -\tilde{\mu}A\mathcal{B}_g\boldsymbol{\phi}_c - \partial_2 A\mathcal{D}_s\boldsymbol{\phi}_c, \tag{6.71b}$$

and the non-resonant terms (N.R.T.) have been omitted. The compatibility condition of Eqs. 6.70, being the operator non-self-adjoint, requires that the know terms are orthogonal (in complex field) to the *left eigenvector* $\boldsymbol{\psi}_c$ of the adjoint problem, associated with the same eigenvalue $i\omega_c$,[23]:

$$\int_\Omega \bar{\boldsymbol{\psi}}_c^T \mathbf{p}_{21}\,\mathrm{d}\Omega + \int_\Gamma \bar{\boldsymbol{\psi}}_c^T \mathbf{q}_{21}\,\mathrm{d}\Gamma = 0, \tag{6.72}$$

from which an ordinary differential equation for $A(t_2)$ is derived. Since $\dot{A} = \epsilon^2\partial_2 A$, by coming back to the unrescaled quantities, it reads:

$$\dot{A} = c_1\tilde{\mu}A + c_3A^2\bar{A}, \tag{6.73}$$

where:

$$\begin{aligned} c_1 :=& -\frac{\int_\Omega \bar{\boldsymbol{\psi}}_c^T\mathcal{K}_g\boldsymbol{\phi}_c\mathrm{d}\Omega + \int_\Gamma \bar{\boldsymbol{\psi}}_c^T\mathcal{B}_g\boldsymbol{\phi}_c\mathrm{d}\Gamma}{2i\omega_c\int_\Omega \bar{\boldsymbol{\psi}}_c^T\mathcal{M}\boldsymbol{\phi}_c\mathrm{d}\Omega + \int_\Omega \bar{\boldsymbol{\psi}}_c^T\mathcal{C}_s\boldsymbol{\phi}_c\mathrm{d}\Omega + \int_\Gamma \bar{\boldsymbol{\psi}}_c^T\mathcal{D}_s\boldsymbol{\phi}_c\mathrm{d}\Gamma}, \\ c_3 :=& -\frac{2\int_\Omega \bar{\boldsymbol{\psi}}_c^T\left[\boldsymbol{n}_2\left(\boldsymbol{\phi}_c, \boldsymbol{\chi}_{10}, \mu_c\right) + \boldsymbol{n}_2\left(\boldsymbol{\phi}_c, \bar{\boldsymbol{\chi}}_{10}, \mu_c\right) + \boldsymbol{n}_2\left(\bar{\boldsymbol{\phi}}_c, \boldsymbol{\chi}_{12}, \mu_c\right)\right]\mathrm{d}\Omega}{2i\omega_c\int_\Omega \bar{\boldsymbol{\psi}}_c^T\mathcal{M}\boldsymbol{\phi}_c\mathrm{d}\Omega + \int_\Omega \bar{\boldsymbol{\psi}}_c^T\mathcal{C}_s\boldsymbol{\phi}_c\mathrm{d}\Omega + \int_\Gamma \bar{\boldsymbol{\psi}}_c^T\mathcal{D}_s\boldsymbol{\phi}_c\mathrm{d}\Gamma} \\ & -\frac{3\int_\Omega \bar{\boldsymbol{\psi}}_c^T\boldsymbol{n}_3\left(\boldsymbol{\phi}_c, \boldsymbol{\phi}_c, \bar{\boldsymbol{\phi}}_c, \mu_c\right)\mathrm{d}\Omega}{2i\omega_c\int_\Omega \bar{\boldsymbol{\psi}}_c^T\mathcal{M}\boldsymbol{\phi}_c\mathrm{d}\Omega + \int_\Omega \bar{\boldsymbol{\psi}}_c^T\mathcal{C}_s\boldsymbol{\phi}_c\mathrm{d}\Omega + \int_\Gamma \bar{\boldsymbol{\psi}}_c^T\mathcal{D}_s\boldsymbol{\phi}_c\mathrm{d}\Gamma}\end{aligned} \tag{6.74}$$

[23] The left eigenvalue satisfies the adjoint of Eqs. 6.66, i.e.:

$$\left(\mathcal{K}_e + \mu_c\mathcal{K}_g^* - i\omega_c\mathcal{C}_s - \omega_c^2\mathcal{M}\right)\boldsymbol{\psi}_c = \mathbf{0},$$

$$\left(\mathcal{B}_s + \mu_0\mathcal{B}_g^* - i\omega_c\mathcal{D}_s\right)\boldsymbol{\psi}_c = \mathbf{0},$$

where $\mathcal{K}_g^*$ is the *adjoint stiffness operator* and $\mathcal{B}_g^*$ the *adjoint boundary conditions*, which are drawn from the Extended Green Identity:

$$\int_\Omega \bar{\boldsymbol{\psi}}^T\mathcal{L}\boldsymbol{\phi}\,\mathrm{d}\Omega + \int_\Gamma \bar{\boldsymbol{\psi}}^T\mathcal{B}\boldsymbol{\phi}\,\mathrm{d}\Gamma = \int_\Omega \bar{\boldsymbol{\phi}}^T\mathcal{L}^*\boldsymbol{\psi}\,\mathrm{d}\Omega + \int_\Gamma \bar{\boldsymbol{\phi}}^T\mathcal{B}^*\boldsymbol{\psi}\,\mathrm{d}\Gamma.$$

From the same identity, the compatibility condition is derived (see the Appendix A, Sect. A.4).

are complex constants. The bifurcation Eq. 6.73 governs the slow motion of amplitude and phase of the leading part of the response, Eq. 6.67.[24] From this one, the limit cycle is computed and its stability ascertained. An example of dynamic bifurcation analysis is carried out in the Problem 7.10.

References

1. Andrianov, I., Awrejcewicz, J., Danishevs'kyy, V., Ivankov, A.: Asymptotic Methods in the Theory of Plates with Mixed Boundary Conditions. Wiley, Chichester (2014)
2. Bolotin, V.V.: Nonconservative Problems of the Theory of Elastic Stability. Macmillan, London (1963)
3. Bolotin, V.V.: The Dynamic Stability of Elastic Systems. Holden Day, San Francisco (1964)
4. Carr, J.: Applications of Centre Manifold Theory. Springer, New York (1981)
5. Guckenheimer, J., Holmes, P.: Nonlinear Oscillations, Dynamical Systems, and Bifurcations of Vector Fields. Springer, New York (1983)
6. Hinch, E.J.: Perturbation Methods. Cambridge University Press, Cambridge (1991)
7. Holmes, M.: Introduction to Perturbation Methods. Springer, New York (2015)
8. Kevorkian, J., Cole, J.D.: Multiple Scale and Singular Perturbation Methods. Springer, New York (2011)
9. Leipholz, H.H.: Stability of Elastic Systems. Sijthoff & Noordhoff, Alphen aan den Rijn (1980)
10. Luongo, A., Ferretti, M., Di Nino, S.: Stability and Bifurcation of Structures: Statical and Dynamical Systems. Springer, Cham (2023)
11. Nayfeh, A.H., Mook, D.T.: Nonlinear Oscillations. Wiley, New York (1995)
12. Nayfeh, A.H.: Perturbation Methods. Wiley, New York (1973)
13. Pignataro, M., Rizzi, N., Luongo, A.: Stability, Bifurcation and Postcritical Behaviour of Elastic Structures. Elsevier, Amsterdam (1990)
14. Rand, R.: Perturbation Methods. Bifurcation Theory and Computer Algebra. Springer, New York (1987)
15. Seyranian, A.P., Mailybaev, A.A.: Multiparameter Stability Theory with Mechanical Applications. World Scientific, Singapore (2003)
16. Thompson, J.M.T., Hunt, G.W.: A General Theory of Elastic Stability. Wiley, London (1973)
17. Thompson, J.M.T., Hunt, G.W.: Elastic Instability Phenomena. Wiley, Chichester (1984)
18. Troger, H., Steindl, A.: Nonlinear Stability and Bifurcation Theory: An Introduction for Engineers and Applied Scientists. Springer, Wien (1991)
19. Wiggins, S.: Introduction to Applied Nonlinear Dynamical Systems and Chaos. Springer, New York (1990)

[24] The motion, therefore, takes place on an (invariant) two dimensional manifold, according to the Center Manifold Theory.

Solved Problems 7

7.1 Introduction

Perturbation methods can be applied, in the same way, to any continuous structure, after having formulated the (integro-) differential equations governing its equilibrium or motion. The algorithms detailed in Chap. 6 for the metamodel are now exemplified for a gallery of 1D and 2D structures, including cables, beams, membranes and plates. The focus is on the analytical procedure leading to build-up the bifurcation equations, which rule the motion of the system on a manifold of reduced dimensions (typically, 1 or 2). After that, the study of such equations closely follows that already illustrated in Chaps. 2–4 for the sample model, and therefore not repeated here. The reader is encouraged to carry out calculations in his own, in order to test his level of learning the theory.

7.2 Nonlinear Elastostatics

A taut string under uniform load (Fig. 7.1a) and a square taut membrane under sinusoidal load (Fig. 7.1b) are considered, and their nonlinear static responses evaluated by a perturbation method. The influence of the nonlinearities is detected.

Problem 7.1 (*Statics of a taut string under uniform load*) Consider a string, fixed at the ends, taut by an axial force N_0, loaded by uniformly distributed transverse forces αp_y (Fig. 7.1a). (a) Write the equilibrium equations in the regime of finite displacements. (b) Solve them by the static perturbation method, and determine the response, commenting the hardening/softening effect of nonlinearities. (c) Evaluate the change of tension induced by the load. (d) To investigate the importance of the nonlinear effects, take $\delta := \frac{p_y \ell}{N_0} = 1$,

A. Luongo et al., *Perturbation Methods and Nonlinear Phenomena*,
Synthesis Lectures on Engineering, Science, and Technology,
https://doi.org/10.1007/978-3-031-49397-3_7

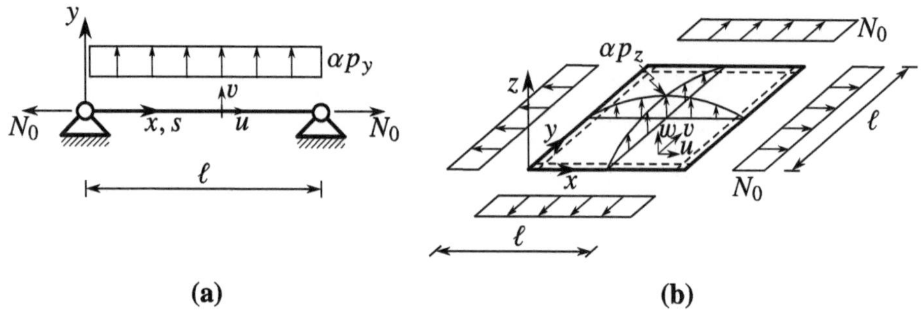

Fig. 7.1 Statics of prestressed structural elements: **a** taut string under uniform load, **b** square taut membrane under sinusoidal load

$\eta := \frac{EA}{N_0} = 500$,[1] and compute the multiplierα for which: (i) the nonlinear deflection at midspan, $d_{NL} := v\left(\frac{\ell}{2}\right)$, is at least 5% different from its linear counterpart d_L; (ii) for this value of α, evaluate the change of tension.

(a) The equilibrium is governed by Eq. 5.7a, with the inertial effects neglected (symbols defined there):

$$-\left(N_0 + \frac{1}{2}\frac{EA}{\ell}\int_0^\ell v'^2 \mathrm{d}s\right) v'' = \alpha p_y, \tag{7.1}$$

to be integrated with the boundary conditions $v(0) = 0$, $v(\ell) = 0$.

(b) Since the system is symmetric, the rescaling $v \to \epsilon^{1/2}\hat{v}$ is performed, together with $\alpha \to \epsilon^{1/2}\hat{\alpha}$; moreover, the expansion (hat omitted) $v = v_0 + \epsilon v_1 + \epsilon^2 v_2 + \cdots$ is introduced. The following perturbation equations are derived[2]:

$$\epsilon^0 : -N_0 v_0'' = \alpha p_y, \tag{7.2a}$$

$$\epsilon^1 : -N_0 v_1'' = \frac{1}{2}\frac{EA}{\ell} v_0'' \int_0^\ell v_0'^2 \mathrm{d}s, \tag{7.2b}$$

$$\epsilon^2 : -N_0 v_2'' = \frac{1}{2}\frac{EA}{\ell}\left(v_1'' \int_0^\ell v_0'^2 \mathrm{d}s + 2v_0'' \int_0^\ell v_0' v_1' \mathrm{d}s\right), \tag{7.2c}$$

---

[1] Since $d_L = \frac{p_y \ell^2}{8N_0}$ is the sag of the string in the linear theory, the parameter used here is also equal to $\delta = \frac{8d_L}{\ell}$; accordingly, $\delta = 1$ denotes a load p_y producing a linear sag $\frac{\ell}{8}$.

[2] For consistency, the series expansion should not exceed the order at which the equation has been truncated. Here, however, we break the rule, due the approximate nature of the minimal model.

together with the boundary conditions $v_k(0) = v_k(\ell) = 0$, $k = 0, 1, \ldots$. By solving the equations in sequence, and reabsorbing ϵ, it follows that:

$$v = \alpha\chi_0(s) + \alpha^3\chi_1(s) + \alpha^5\chi_2(s) + \cdots, \tag{7.3}$$

with:

$$\chi_0 := \frac{1}{2}\delta\left(1 - \frac{s}{\ell}\right)s, \quad \chi_1 := -\frac{1}{48}\eta\delta^3\left(1 - \frac{s}{\ell}\right)s, \quad \chi_2 := \frac{1}{384}\eta^2\delta^5\left(1 - \frac{s}{\ell}\right)s \tag{7.4}$$

and $\delta := \frac{p_y\ell}{N_0}$, $\eta := \frac{EA}{N_0}$.

Cubic nonlinearities produce a hardening effect, reducing the linear deflection; conversely, quintic nonlinearities produce a softening effect.

(c) The change of tension, uniform along the string, is given by Eq. 5.6. By dividing it by N_0, and using Eq. 7.3, the relative change of tension is derived, i.e.:

$$\frac{\tilde{N}}{N_0} = \frac{EA}{2N_0\ell}\int_0^\ell v'^2\mathrm{d}s = \frac{\eta}{2\ell}\int_0^\ell\left(\alpha^2\chi_0'^2 + 2\alpha^4\chi_0'\chi_1' + \cdots\right)\mathrm{d}s = \frac{\eta\alpha^2\delta^2}{24}\left(1 - \frac{1}{12}\alpha^2\delta^2\eta\right), \tag{7.5}$$

where the series has been truncated for consistency.

(d) By taking $\delta = 1$, the linear and nonlinear deflections at midspan are evaluated by Eq. 7.3, namely $d_L = \alpha\chi_0\left(\frac{\ell}{2}\right) = \alpha\frac{\ell}{8}$ (see the Note 1), and

$$d_{NL} = \alpha\chi_0\left(\frac{\ell}{2}\right) + \alpha^3\chi_1\left(\frac{\ell}{2}\right) + \alpha^5\chi_2\left(\frac{\ell}{2}\right) + \cdots = \alpha\frac{\ell}{8} - \alpha^3\frac{\eta\ell}{192} + \alpha^5\frac{\eta^2\ell}{1536}, \tag{7.6}$$

which depends on the stiffness-to-tension ratio η. By taking, e.g., $\eta = 500$, it is $\frac{d_{NL}}{d_L} = 1 - \frac{125}{6}\alpha^2 + \frac{15625}{12}\alpha^4$. By requiring this ratio is equal to 0.95, $\alpha = 0.0542$ is found. It means that the load p_y that produces a deflection $d_L = \frac{\ell}{8}$ in the linear field, must be amplified by 0.0542, in order to trigger an appreciable deviation of 5% from linearity. For this value of α, using Eq. 7.5, the relative tension is found to be $\frac{\tilde{N}}{N_0} = 0.0538$. □

Problem 7.2 (*Statics of a square membrane under sinusoidal load*) Consider a square membrane of side ℓ, lying in the (x, y) plane, uniformly taut by a hydrostatic planar tension N_0, subjected to transverse loads $\alpha\breve{p}_z\sin\left(\frac{\pi x}{\ell}\right)\sin\left(\frac{\pi y}{\ell}\right)$ (Fig. 7.1b). The out-of-plane displacements and the in-plane displacements normal to the boundary are prevented; tangential displacements are free. (a) Write the equilibrium equations in the regime of finite displacements. (b) Solve them by the static perturbation method and determine the response. (c) Find the change of tension induced by the load, and evaluate it at midpoint for significant values of the membrane stiffness and deflection.

(a) The equilibrium is governed by Eqs. 5.24 and 5.25, in which the inertia forces are ignored, i.e.:

$$
\begin{aligned}
&-N_0(w_{,xx}+w_{,yy})-C\Big[u_{,x}\,w_{,xx}+v_{,y}\,w_{,yy}+\nu u_{,x}\,w_{,yy}+\nu v_{,y}\,w_{,xx}+(1-\nu)u_{,y}\,w_{,xy}\\
&+(1-\nu)v_{,x}\,w_{,xy}\Big]-C\Big[\frac{1}{2}w_{,x}^2\,w_{,xx}+\frac{1}{2}w_{,y}^2\,w_{,yy}+\frac{1}{2}\nu w_{,y}^2\,w_{,xx}+\frac{1}{2}\nu\, w_{,x}^2\,w_{,yy}\\
&+(1-\nu)w_{,x}\,w_{,y}\,w_{,xy}\Big]=\alpha\breve{p}_z\sin\left(\frac{\pi x}{\ell}\right)\sin\left(\frac{\pi y}{\ell}\right),
\end{aligned}
\tag{7.7}
$$

and

$$u_{,xx}+\check{\nu}\,u_{,yy}+\hat{\nu}\,v_{,xy}+w_{,x}\,w_{,xx}+\check{\nu}w_{,x}\,w_{,yy}+\hat{\nu}\,w_{,y}\,w_{,xy}=0, \tag{7.8a}$$

$$\hat{\nu}\,u_{,xy}+v_{yy}+\check{\nu}\,v_{,xx}+\check{\nu}\,w_{,y}\,w_{,xx}+w_{,y}\,w_{,yy}+\hat{\nu}\,w_{,x}\,w_{,xy}=0, \tag{7.8b}$$

where $\hat{\nu}:=\frac{1+\nu}{2}$ and $\check{\nu}:=\frac{1-\nu}{2}$; moreover, the boundary conditions are:

$$u=0, \qquad \text{at } x=0,\ \ell, \tag{7.9a}$$

$$v=0, \qquad \text{at } y=0,\ \ell, \tag{7.9b}$$

$$\tilde{N}_{xy}=C\check{\nu}\left(u_{,y}+v_{,x}+w_{,x}\,w_{,y}\right)=0, \qquad \text{at } x,\,y=0,\ \ell, \tag{7.9c}$$

$$w=0, \qquad \text{at } x,\,y=0,\ \ell. \tag{7.9d}$$

(b) Due to the presence of quadratic nonlinearities, displacements are rescaled as $(u,v,w)\to\epsilon\left(\hat{u},\hat{v},\hat{w}\right)$, and the load as $\alpha\to\epsilon\hat{\alpha}$. Moreover, even/odd series are introduced (hat omitted)[3]:

$$w=w_0+\epsilon^2 w_2+\cdots, \tag{7.10a}$$

$$u=\epsilon u_1+\cdots, \tag{7.10b}$$

$$v=\epsilon v_1+\cdots. \tag{7.10c}$$

The following perturbation equations are derived:

Order ϵ^0:

$$-N_0\left(w_{0,xx}+w_{0,yy}\right)=\alpha\breve{p}_z\sin\left(\frac{\pi x}{\ell}\right)\sin\left(\frac{\pi y}{\ell}\right). \tag{7.11}$$

Order ϵ^1:

$$u_{1,xx}+\check{\nu}u_{1,yy}+\hat{\nu}v_{1,xy}=-w_{0,x}w_{0,xx}-\check{\nu}w_{0,x}w_{0,yy}-\hat{\nu}w_{0,y}w_{0,xy}, \tag{7.12a}$$

$$\hat{\nu}u_{1,xy}+v_{1,yy}+\check{\nu}v_{1,xx}=-\check{\nu}w_{0,y}w_{0,xx}-w_{0,y}w_{0,yy}-\hat{\nu}w_{0,x}w_{0,xy}. \tag{7.12b}$$

---

[3] Indeed, by changing the sign of ϵ, $\epsilon\hat{w}$ changes sign, while $\epsilon\left(\hat{u},\hat{v}\right)$ do not, according to the physics of the problem.

Order ϵ^2:

$$\begin{aligned}
-N_0(w_2,_{xx} + w_2,_{yy}) = & C\Big[u_1,_x\, w_0,_{xx} + v_1,_y\, w_0,_{yy} + \nu u_1,_x\, w_0,_{yy} + \nu v_1,_y\, w_0,_{xx} \\
& + (1-\nu) u_1,_y\, w_0,_{xy} + (1-\nu) v_1,_x\, w_0,_{xy}\Big] + C\Big[\frac{1}{2} w_0,_x^2\, w_0,_{xx} \\
& + \frac{1}{2} w_0,_y^2\, w_0,_{yy} + \frac{1}{2}\nu w_0,_y^2\, w_0,_{xx} + \frac{1}{2}\nu\, w_0,_x^2\, w_0,_{yy} \\
& + (1-\nu) w_0,_x\, w_0,_y\, w_0,_{xy}\Big],
\end{aligned} \tag{7.13}$$

together with the boundary conditions:

$$w_k = 0, \quad \text{at } x, y = 0,\ \ell, \quad k = 0, 2, \tag{7.14}$$

and

$$u_1 = 0, \qquad \text{at } x = 0,\ \ell, \tag{7.15a}$$

$$v_1 = 0, \qquad \text{at } y = 0,\ \ell, \tag{7.15b}$$

$$C\check{\nu}\left(u_1,_y + v_1,_x\right) = -C\check{\nu} w_0,_x\, w_0,_y\,, \qquad \text{at } x, y = 0,\ \ell. \tag{7.15c}$$

By solving the equations in chain, it follows that:

$$w = \alpha W_0(x, y) + \alpha^3 W_2(x, y) + \cdots, \tag{7.16a}$$

$$u = \alpha^2 U_1(x, y) + \cdots, \tag{7.16b}$$

$$v = \alpha^2 V_1(x, y) + \cdots, \tag{7.16c}$$

where:

$$W_0 := \breve{w} \sin\left(\frac{\pi x}{\ell}\right) \sin\left(\frac{\pi y}{\ell}\right), \tag{7.17a}$$

$$U_1 := \breve{w}^2 \frac{\pi}{16\,\ell}\left[\cos\left(\frac{2\pi y}{\ell}\right) - 2\check{\nu}\right] \sin\left(\frac{2\pi x}{\ell}\right), \tag{7.17b}$$

$$V_1 := \breve{w}^2 \frac{\pi}{16\,\ell}\left[\cos\left(\frac{2\pi x}{\ell}\right) - 2\check{\nu}\right] \sin\left(\frac{2\pi y}{\ell}\right), \tag{7.17c}$$

$$W_2 := \frac{C}{N_0} \breve{w}^3 \frac{\pi^2}{40\ell^2} \hat{\nu}\left[2\check{\nu} \cos\left(\frac{2\pi x}{\ell}\right) + 2\check{\nu} \cos\left(\frac{2\pi y}{\ell}\right) - 8\check{\nu} - 10\right] \sin\left(\frac{\pi x}{\ell}\right) \sin\left(\frac{\pi y}{\ell}\right), \tag{7.17d}$$

in which $\breve{w} := \frac{\breve{p}_z \ell^2}{2N_0\pi^2}$ is the deflection at midpoint when $\alpha = 1$, according to the linear theory.

(c) The incremental stress components, truncated for consistency, are evaluated by Eqs. 5.22, 5.23, and by using Eqs. 7.16, i.e.:

$$\tilde{N}_x = C\alpha^2 \left[U_{1,x} + \frac{1}{2} W_{0,x}^2 + \nu \left(V_{1,y} + \frac{1}{2} W_{0,y}^2 \right) + \cdots \right] = C \left(\frac{\alpha\pi\breve{w}}{2\ell} \right)^2 \hat{\nu} \left[1 - 2\breve{\nu} \cos\left(\frac{2\pi y}{\ell} \right) \right], \tag{7.18a}$$

$$\tilde{N}_y = C\alpha^2 \left[\nu \left(U_{1,x} + \frac{1}{2} W_{0,x}^2 \right) + V_{1,y} + \frac{1}{2} W_{0,y}^2 + \cdots \right] = C \left(\frac{\alpha\pi\breve{w}}{2\ell} \right)^2 \hat{\nu} \left[1 - 2\breve{\nu} \cos\left(\frac{2\pi x}{\ell} \right) \right], \tag{7.18b}$$

$$\tilde{N}_{xy} = C\breve{\nu}\alpha^2 \left[U_{1,y} + V_{1,x} + W_{0,x} W_{0,y} + \cdots \right] = 0. \tag{7.18c}$$

It is important to note that due to the symmetry of the structure and the loads, $\tilde{N}_{xy} = 0$. Therefore, in the case under examination, the membrane behaves like two orders of orthogonal taut strings.

By evaluating the stresses at the midpoint $(x, y) = \left(\frac{\ell}{2}, \frac{\ell}{2}\right)$, and dividing them by the static tension N_0, $\left[\frac{\tilde{N}_x}{N_0}, \frac{\tilde{N}_y}{N_0}, \frac{\tilde{N}_{xy}}{N_0} \right]_{x=\frac{\ell}{2}, y=\frac{\ell}{2}} = \frac{C}{N_0} \left(\frac{\alpha\pi\breve{w}}{2\ell} \right)^2 \hat{\nu} \left(1 + 2\breve{\nu} \right) [1, 1, 0]$ follows. By taking $\frac{C}{N_0} = 500$ and $\alpha\frac{\breve{w}}{\ell} = \frac{1}{10}$, we get: $\left[\frac{\tilde{N}_x}{N_0}, \frac{\tilde{N}_y}{N_0}, \frac{\tilde{N}_{xy}}{N_0} \right]_{x=\frac{\ell}{2}, y=\frac{\ell}{2}} = \frac{5\pi^2}{4} \hat{\nu} \left(1 + 2\breve{\nu} \right) [1, 1, 0]$. Therefore, it needs to take $\alpha\frac{\breve{w}}{\ell} = \frac{1}{25\pi\sqrt{2\hat{\nu}(1+2\breve{\nu})}}$ to obtain an incremental tension of the order of 10% of the static one. □

7.3 Buckling and Postbuckling

A rectangular plate, uniformly compressed in one direction, is considered, and its buckling and postbuckling investigated (Fig. 7.2).

Problem 7.3 (*Buckling and postbuckling of a rectangular hinged plate, uniformly compressed in one direction*) Consider a rectangular plate, occupying the domain $\Omega := \{(x, y) \mid 0 \le x \le \ell, 0 \le y \le b\}$, with integer aspect ratio $\frac{\ell}{b}$, hinged at the boundary Γ. In-plane displacements normal to the boundary are prevented, while tangential displacements are free. The plate is uniformly compressed by longitudinal forces per unit length

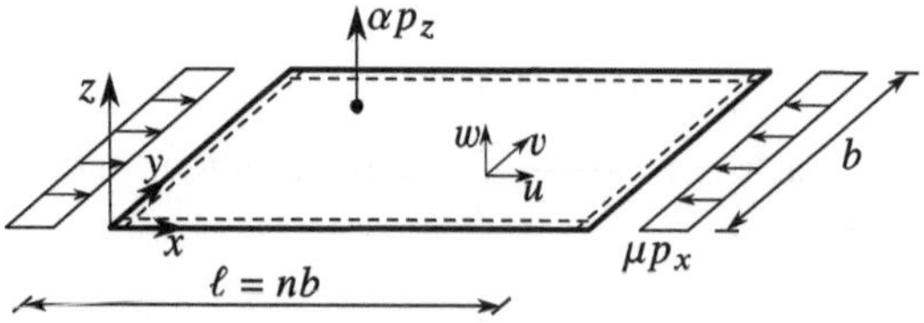

Fig. 7.2 Buckling of a rectangular plate, with integer aspect ratio, hinged at the boundary and uniformly compressed in the longitudinal direction

$\mu p_x = \text{const}$,[4] applied at the sides $x = 0, \ell$, and loaded by small constant transverse force αp_z (Fig. 7.2). (a) Write the equilibrium equations in the regime of finite displacements. (b) Derive the perturbation equations. (c) Evaluate the critical value μ_c of the load multiplier. (d) Determine the bifurcated path in absence or in presence of transverse forces.

(a) The out-of-plane balance equations of the plate are drawn from Eq. 5.28. Particularizing it to the static regime and by letting $N_x^0 = -\mu p_x$, $N_y^0 = N_{xy}^0 = 0$, it follows:

$$\begin{aligned} &D\nabla^4 w + \mu p_x w_{,xx} - C\Big[u_{,x}\, w_{,xx} + v_{,y}\, w_{,yy} + \nu\, u_{,x}\, w_{,yy} + \nu\, v_{,y}\, w_{,xx} \\ &+ (1-\nu) u_{,y}\, w_{,xy} + (1-\nu) v_{,x}\, w_{,xy}\Big] - C\Big[\frac{1}{2} w_{,x}^2\, w_{,xx} + \frac{1}{2} w_{,y}^2\, w_{,yy} \\ &+ \frac{1}{2}\nu\, w_{,y}^2\, w_{,xx} + \frac{1}{2}\nu\, w_{,x}^2\, w_{,yy} + (1-\nu)\, w_{,x}\, w_{,y}\, w_{,xy}\Big] = \alpha p_z, \end{aligned} \tag{7.19}$$

The in-plane balance equations are given by Eqs. 7.8. The boundary conditions prescribe: (i) the vanishing of the normal displacements and of the tangent stress at the boundary Γ, i.e.,

$$u = 0, \qquad \text{at } x = 0,\ \ell, \tag{7.20a}$$

$$v = 0, \qquad \text{at } y = 0,\ b, \tag{7.20b}$$

$$\tilde{N}_{xy} = C\check{\nu}\left(u_{,y} + v_{,x} + w_{,x}\, w_{,y}\right) = 0, \qquad \text{at } x = 0,\ \ell,\ \ y = 0,\ b, \tag{7.20c}$$

$$w = 0, \qquad \text{at } x = 0,\ \ell,\ \ y = 0,\ b; \tag{7.20d}$$

(ii) the vanishing of the second normal derivative of the transverse displacement, i.e., $w_{,xx} = 0$ at $x = 0, \ell$ and $w_{,yy} = 0$ at $y = 0, b$.

(b) The following rescaling is introduced: $(u, v, w) \to \epsilon\left(\hat{u}, \hat{v}, \hat{w}\right)$, and $\alpha \to \epsilon^3\hat{\alpha}$. Then, the series in Eq. 7.10 are introduced for displacements; moreover, the bifurcation parameter is expanded as:

$$\mu = \mu_0 + \epsilon^2 \mu_2 + \cdots. \tag{7.21}$$

The following perturbation equations are derived:

Order ϵ^0:

$$D\nabla^4 w_0 + \mu_0 p_x w_{0,xx} = 0. \tag{7.22}$$

Order ϵ^1:

$$u_{1,xx} + \check{\nu} u_{1,yy} + \hat{\nu} v_{1,xy} = -w_{0,x} w_{0,xx} - \check{\nu} w_{0,x} w_{0,yy} - \hat{\nu} w_{0,y} w_{0,xy}, \tag{7.23a}$$

$$\hat{\nu} u_{1,xy} + v_{1,yy} + \check{\nu} v_{1,xx} = -\check{\nu} w_{0,y} w_{0,xx} - w_{0,y} w_{0,yy} - \hat{\nu} w_{0,x} w_{0,xy}. \tag{7.23b}$$

[4] In this chapter, the symbol μ will be used to denote any load multipliers, to avoid confusion with the Poisson ratio ν.

Order ϵ^2:

$$D\nabla^4 w_2 + \mu_0 p_x w_{2,xx} = -\mu_2 p_x w_{0,xx} + C\Big[u_{1,x}\, w_{0,xx} + v_{1,y}\, w_{0,yy} + \nu\, u_{1,x}\, w_{0,yy}$$
$$+ \nu\, v_{1,y}\, w_{0,xx} + (1-\nu)u_{1,y}\, w_{0,xy} + (1-\nu)v_{1,x}\, w_{0,xy}\Big]$$
$$+ C\Big[\frac{1}{2} w_{0,x}^{\,2}\, w_{0,xx} + \frac{1}{2} w_{0,y}^{\,2}\, w_{0,yy} + \frac{1}{2}\nu\, w_{0,y}^{\,2}\, w_{0,xx}$$
$$+ \frac{1}{2}\nu\, w_{0,x}^{\,2}\, w_{0,yy} + (1-\nu)\, w_{0,x}\, w_{0,y}\, w_{0,xy}\Big] + \alpha p_z, \tag{7.24}$$

together with the boundary conditions on Γ:

$$w_k = 0, \qquad \text{at } x = 0,\ \ell,\ \ y = 0,\ b, \tag{7.25a}$$
$$w_{k,xx} = 0 \qquad \text{at } x = 0,\ \ell, \qquad k = 0, 2, \tag{7.25b}$$
$$w_{k,yy} = 0 \qquad \text{at } y = 0,\ b, \tag{7.25c}$$

and

$$u_1 = 0, \qquad \text{at } x = 0,\ \ell, \tag{7.26a}$$
$$v_1 = 0, \qquad \text{at } y = 0,\ b, \tag{7.26b}$$
$$C\check{\nu}\left(u_{1,y} + v_{1,x}\right) = -C\check{\nu} w_{0,x}\, w_{0,y}\,, \qquad \text{at } x = 0,\ \ell,\ \ y = 0,\ b. \tag{7.26c}$$

To normalize the solution, $w\left(\frac{b}{2}, \frac{b}{2}\right) = a$ is used, with a the amplitude; consequently, $w_0\left(\frac{b}{2}, \frac{b}{2}\right) = a$ and $w_k\left(\frac{b}{2}, \frac{b}{2}\right) = 0,\ k = 2, 4, \ldots$.

(c) The ϵ^0 problem admits the normalized solution $w_0 = a \sin\left(\frac{n\pi x}{\ell}\right) \sin\left(\frac{m\pi y}{b}\right)$, with n, m arbitrary integers; the associated ∞^2 critical loads are:

$$\mu_0 = \frac{\pi^2 D}{p_x b^2}\left(\frac{nb}{\ell} + m^2\frac{\ell}{nb}\right)^2. \tag{7.27}$$

The minimum of them, $\mu_c := \min\limits_{n,m} \mu_0$, is found for $m = 1$, and, since $\frac{\ell}{b}$ is an integer, for $n = \frac{\ell}{b}$,[5]; hence, $\mu_c = 4\frac{\pi^2 D}{p_x b^2}$. Correspondingly, the critical mode $w_0 = w_c$ is:

$$w_c = a \sin\left(\frac{\pi x}{b}\right) \sin\left(\frac{\pi y}{b}\right) =: a\phi_c. \tag{7.28}$$

i.e., the plate buckles in square fields.

[5] If the aspect ratio were not an integer, the integer n closest to the ratio should be taken, resulting in a slight higher critical load (see, e.g., [2, 4]).

(d) To analyze the postcritical behavior, the higher order perturbation equations must be solved. Equations 7.23 and 7.26 read[6]:

$$u_{1,xx} + \check{\nu} u_{1,yy} + \hat{\nu} v_{1,xy} = -a^2 \left(\phi_{c,x}\, \phi_{c,xx} + \check{\nu} \phi_{c,x}\, \phi_{c,yy} + \hat{\nu}\, \phi_{c,y}\, \phi_{c,xy}\right), \tag{7.29a}$$

$$\hat{\nu} u_{1,xy} + v_{1,yy} + \check{\nu} v_{1,xx} = -a^2 \left(\check{\nu}\, \phi_{c,y}\, \phi_{c,xx} + \phi_{c,y}\, \phi_{c,yy} + \hat{\nu}\, \phi_{c,x}\, \phi_{c,xy}\right), \tag{7.29b}$$

$$u_1 = 0, \qquad \text{at } x = 0,\ \ell, \tag{7.30a}$$

$$v_1 = 0, \qquad \text{at } y = 0,\ b, \tag{7.30b}$$

$$C\check{\nu} \left(u_{1,y} + v_{1,x}\right) = 0, \qquad \text{at } x = 0,\ \ell,\ \ y = 0,\ b, \tag{7.30c}$$

whose solution is:

$$u_1 = a^2 U_1 (x, y), \tag{7.31a}$$

$$v_1 = a^2 V_1 (x, y), \tag{7.31b}$$

where:

$$U_1 := \frac{\pi}{16b} \left[\cos\left(\frac{2\pi y}{b}\right) - 2\check{\nu}\right] \sin\left(\frac{2\pi x}{b}\right), \tag{7.32a}$$

$$V_1 := \frac{\pi}{16b} \left[\cos\left(\frac{2\pi x}{b}\right) - 2\check{\nu}\right] \sin\left(\frac{2\pi y}{b}\right). \tag{7.32b}$$

By substituting them in Eq. 7.24, this latter becomes:

$$\begin{aligned} D\nabla^4 w_2 + \mu_c p_x w_{2,xx} = & -a\mu_2 p_x \phi_{c,xx} + a^3 C\Big[U_{1,x}\, \phi_{c,xx} + V_{1,y}\, \phi_{c,yy} + \nu\, U_{1,x}\, \phi_{c,yy} \\ & + \nu\, V_{1,y}\, \phi_{c,xx} + (1-\nu) U_{1,y}\, \phi_{c,xy} + (1-\nu) V_{1,x}\, \phi_{c,xy}\Big] \\ & + a^3 C\Big[\frac{1}{2}\phi_{c,x}^2\, \phi_{c,xx} + \frac{1}{2}\phi_{c,y}^2\, \phi_{c,yy} + \frac{1}{2}\nu\, \phi_{c,y}^2\, \phi_{c,xx} \\ & + \frac{1}{2}\nu\, \phi_{c,x}^2\, \phi_{c,yy} + (1-\nu)\, \phi_{c,x}\, \phi_{c,y}\, \phi_{c,xy}\Big] + \alpha p_z. \end{aligned} \tag{7.33}$$

Compatibility requires the known term is orthogonal to ϕ_c, from which μ_2 is derived as:

$$\mu_2 = C_2 a^2 + \frac{\alpha}{a} C_\alpha, \tag{7.34}$$

[6] It should be noticed, that these equations are non-singular, so that their solution is unique.

where:

$$C_2 := \frac{C}{p_x \int_0^{nb} \int_0^b \phi_c \, \phi_{c,xx} \, \mathrm{d}x\mathrm{d}y} \int_0^{nb} \int_0^b \phi_c \Big[U_{1,x} \, \phi_{c,xx} + V_{1,y} \, \phi_{c,yy} \tag{7.35a}$$

$$+ \nu \, U_{1,x} \, \phi_{c,yy} + \nu \, V_{1,y} \, \phi_{c,xx} + (1-\nu) U_{1,y} \, \phi_{c,xy}$$

$$+ (1-\nu) V_{1,x} \, \phi_{c,xy} + \frac{1}{2} \phi_{c,x}^2 \, \phi_{c,xx} + \frac{1}{2} \phi_{c,y}^2 \, \phi_{c,yy} + \frac{1}{2} \nu \, \phi_{c,y}^2 \, \phi_{c,xx}$$

$$+ \frac{1}{2} \nu \, \phi_{c,x}^2 \, \phi_{c,yy} + (1-\nu) \, \phi_{c,x} \, \phi_{c,y} \, \phi_{c,xy} \Big] \mathrm{d}x\mathrm{d}y = \frac{\pi^2 C}{4 b^2 p_x} (1+\nu) \left(1 + \check{\nu}\right),$$

$$C_\alpha := \frac{p_z}{p_x \int_0^{nb} \int_0^b \phi_c \, \phi_{c,xx} \, \mathrm{d}x\mathrm{d}y} \int_0^{nb} \int_0^b \phi_c \mathrm{d}x\mathrm{d}y = \frac{8 b^2 p_z}{\pi^4 p_x} \frac{(-1)^n - 1}{n}. \tag{7.35b}$$

By coming back to the unrescaled quantities, the amplitude-load relationship is finally found:

$$\mu = \mu_c + C_2 a^2 + \frac{\alpha}{a} C_\alpha. \tag{7.36}$$

When the transverse forces are absent ($\alpha = 0$), since $C_2 > 0$, a supercritical pitchfork is found, as represented in Fig. 2.2a. When the transverse forces are present ($\alpha \neq 0$), non-trivial paths, as those represented in Fig. 2.3c, are determined. □

7.4 Free Vibrations and External Resonances

The nonlinear free vibrations of a flat suspended cable are analyzed (Fig. 7.3a). Then, the primary external resonance problem is addressed, referred to: (i) an axially unrestrained beam undergoing prescribed harmonic motion at support (Fig. 7.3b), and (ii) a square plate under harmonic transverse load (Fig. 7.3c). Successively, sub-harmonic and super-harmonic resonances of a longitudinally restrained beam, induced by transverse loads, are studied (Fig. 7.3d). Finally, a comparison between the nonlinear behavior of axially unrestrained and restrained beams is performed.

Problem 7.4 (*Free vibrations of a suspended cable*) Consider a suspended cable, fixed at the ends placed at the same level, having small sag-to-span ratio d/ℓ and subject to self-weight $p_y = -mg$ (Fig. 7.3a). (a) Write the equation which govern the free undamped motion. (b) Apply the Multiple Scale Method (MSM) to derive the perturbation equations. (c) Determine the linear normal modes, antisymmetric and symmetric. (d) Obtain the amplitude-frequency relationship for the fundamental mode, of symmetric or antisymmetric type. (e) Comment the relative importance of quadratic and cubic nonlinearities. (f) Particularize the relationship to a taut string.

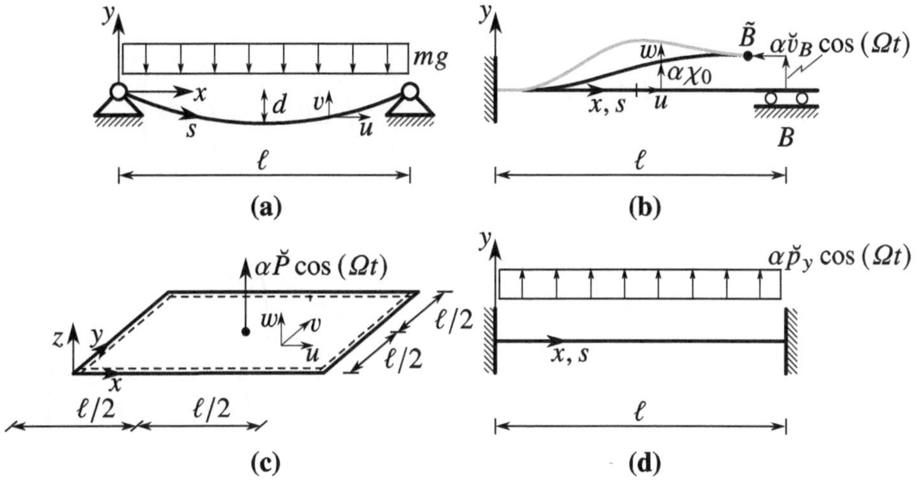

Fig. 7.3 Free and forced dynamics of structural elements: **a** suspended flat cable under self-weight; **b** clamped-sliding beam under resonant harmonic motion at a support; **c** hinged rectangular plate, loaded at the midpoint by a resonant harmonic force; **d** clamped-clamped beam under harmonic uniformly distributed load, in sub- or super-harmonic resonance

(a) The equations of the free motion of a flat suspended cable are given by Eq. 5.10a, with the incremental external forces ignored, i.e.:

$$m\ddot{v} - N_0 v'' + \frac{EA}{\ell}k_0^2 \int_0^\ell v\mathrm{d}s - \frac{EA}{2\ell}k_0 \int_0^\ell v'^2\mathrm{d}s + \frac{EA}{\ell}k_0 v'' \int_0^\ell v\mathrm{d}s - \frac{EA}{2\ell}v'' \int_0^\ell v'^2\mathrm{d}s = 0, \tag{7.37}$$

combined to the boundary conditions $v(0,t) = v(\ell,t) = 0$. Here $N_0 := \frac{mg\ell^2}{8d}$ is the static tension and $k_0 := \frac{mg}{N_0} = \frac{8d}{\ell^2}$ is the static curvature, both assumed constant.

(b) By performing the rescaling $v \to \epsilon v$ and dividing by ϵ, the quadratic nonlinearities are shifted to the ϵ-order, and the cubic nonlinearities to the ϵ^2-order. Independent time scales $t_0 := t$, $t_2 := \epsilon^2 t, \cdots$ are introduced, so that $\ddot{v} = \left(\partial_0^2 + 2\partial_0\partial_2 + \cdots\right) v$. Finally, the displacement is expanded as $v = v_0 + \epsilon v_1 + \epsilon^2 v_2 + \cdots$. The following perturbation equations are drawn:

Order ϵ^0:

$$m\partial_0^2 v_0 - N_0 v_0'' + \frac{EA}{\ell}k_0^2 \int_0^\ell v_0\mathrm{d}s = 0. \tag{7.38}$$

Order ϵ^1:

$$m\partial_0^2 v_1 - N_0 v_1'' + \frac{EA}{\ell} k_0^2 \int_0^\ell v_1 \mathrm{d}s = \frac{EA}{\ell} k_0 \left(\frac{1}{2} \int_0^\ell v_0'^2 \mathrm{d}s - v_0'' \int_0^\ell v_0 \mathrm{d}s \right). \tag{7.39}$$

Order ϵ^2:

$$\begin{aligned} m\partial_0^2 v_2 - N_0 v_2'' + \frac{EA}{\ell} k_0^2 \int_0^\ell v_2 \mathrm{d}s &= -2m\partial_0\partial_2 v_0 + \frac{EA}{2\ell} v_0'' \int_0^\ell v_0'^2 \mathrm{d}s \\ &+ \frac{EA}{\ell} k_0 \left(\int_0^\ell v_0' v_1' \mathrm{d}s - v_0'' \int_0^\ell v_1 \mathrm{d}s - v_1'' \int_0^\ell v_0 \mathrm{d}s \right), \end{aligned} \tag{7.40}$$

with the boundary conditions $v_k(0, t_k) = v_k(\ell, t_k) = 0$, $k = 0, 1, \ldots$. As a normalization condition, it is required that the ω_r-harmonic component of the response $v(s, t)$, evaluated at $s = s^*$, with s^* arbitrarily selected, is equal to $a_r(t)$.

(c) The ϵ^0 order perturbation Eq. 7.38, with relevant boundary conditions, admits the following monomodal solution:

$$v_0 = A_r(t_1, \ldots)\, \phi_r(s)\, e^{i\omega_r t_0} + \text{c.c.}, \tag{7.41}$$

in which (ω_r, ϕ_r) is the rth eigenpair. Two cases are distinguished:

Antisymmetric modes. It is $\omega_r = \frac{2r\pi}{\ell}\sqrt{\frac{N_0}{m}}$, $r = 1, 2, \ldots$, with the associated normalized mode:

$$\phi_r = \sin\left(\frac{2r\pi}{\ell} s\right); \tag{7.42}$$

Symmetric modes. It is $\omega_r = \beta_r \sqrt{\frac{N_0}{m}}$, with the wave-number β_r a root of the following transcendental equation (see, e.g., [1]):

$$\tan\left(\frac{\beta_r \ell}{2}\right) = \left(\frac{\beta_r \ell}{2}\right) - \frac{4}{\lambda^2}\left(\frac{\beta_r \ell}{2}\right)^3, \tag{7.43}$$

in which $\lambda^2 := \frac{EA}{N_0}(k_0 \ell)^2$ is a nondimensional elasto-geometric parameter, known as the 'Irvine parameter'. The associated normalized mode, is:

$$\phi_r = \alpha_r[\cos(\beta_r s) - 1 + b_r \sin(\beta_r s)]. \tag{7.44}$$

with $b_r := \tan\left(\frac{\beta_r \ell}{2}\right)$ and the normalization factor $\alpha_r := \frac{1}{\cos\left(\frac{\beta_r \ell}{2}\right) - 1 + b_r \sin\left(\frac{\beta_r \ell}{2}\right)}$.

When $\lambda^2 < 4\pi^2$ the fundamental mode is symmetric; when $\lambda^2 > 4\pi^2$ is antisymmetric [1].

(d) With Eq. 7.41, the ϵ order perturbation Eq. 7.39 becomes:

$$m\partial_0^2 v_1 - N_0 v_1'' + \frac{EA}{\ell} k_0^2 \int_0^\ell v_1 \mathrm{d}s = \frac{EA}{\ell} k_0 \left(\frac{1}{2} \int_0^\ell \phi_r'^2 \mathrm{d}s - \phi_r'' \int_0^\ell \phi_r \mathrm{d}s \right) \left(A_r^2 e^{2i\omega_r t_0} + A_r \bar{A}_r \right) + \text{c.c.}. \tag{7.45}$$

This equation does not require solvability, since the know terms are, either: (a) nonresonant in the symmetric case, or (b) resonant in the antisymmetric case, but they are orthogonal to ϕ_r. When solved, with the relevant boundary conditions, the equation supplies:

$$v_1 = A_r^2 \chi_{12}(s)\, e^{2i\omega_r t_0} + A_r \bar{A}_r \chi_{10}(s) + \text{c.c.}, \tag{7.46}$$

where:

$$\chi_{12} = \begin{cases} \dfrac{\cos\left(\frac{4r\pi s}{\ell}\right) - 1}{k_0 \ell^2 \left[\left(\frac{4}{\lambda}\right)^2 - \frac{1}{(r\pi)^2} \right]}, & \text{antisym}, \\ \dfrac{\alpha_r^2 \beta_r^2}{4k_0} \dfrac{\frac{\beta_r \ell - \sin(\beta_r \ell)}{1+\cos(\beta_r \ell)} - \left(\frac{4}{\lambda}\right)^2 \left(\frac{\beta_r \ell}{2}\right)^3}{\left(\frac{4}{\lambda}\right)^2 \left(\frac{\beta_r \ell}{2}\right)^3 + \frac{\tan(\beta_r \ell)}{2} - \frac{\beta_r \ell}{2}} [\cos(2\beta_r s) - 1 & \text{sym}, \\ + \tan(\beta_r \ell) \sin(2\beta_r s)] + \dfrac{\alpha_r \beta_r^2}{3k_0} \phi_r, & \end{cases} \tag{7.47}$$

$$\chi_{10} = \begin{cases} \dfrac{6 (r\pi)^2}{k_0 \ell^4 \left(1 + \frac{12}{\lambda^2}\right)} (\ell - s) s, & \text{antisym}, \\ \dfrac{\alpha_r \beta_r}{k_0} \left[\dfrac{3\alpha_r}{\ell^3 \left(1 + \frac{12}{\lambda^2}\right)} \left(\dfrac{\beta_r \ell - \sin(\beta_r \ell)}{1 + \cos(\beta_r \ell)} - \left(\dfrac{4}{\lambda}\right)^2 \left(\dfrac{\beta_r \ell}{2}\right)^3 \right) (\ell - s) s - \beta_r \phi_r \right], & \text{sym}. \end{cases} \tag{7.48}$$

With the previous results the ϵ^2-order perturbation Eq. 7.40 reads:

$$m\partial_0^2 v_2 - N_0 v_2'' + \frac{EA}{\ell} k_0^2 \int_0^\ell v_2 \mathrm{d}s = q_{21}(s) e^{i\omega_r t_0} + \text{N.R.T.} + \text{c.c.}, \tag{7.49}$$

where:

$$\begin{aligned} q_{21} := & -2im\omega_r \phi_r \partial_2 A_r + \frac{EA}{\ell} k_0 A_r^2 \bar{A}_r \Big[\int_0^\ell \phi_r' \left(\chi_{12}' + 2\chi_{10}'\right) \mathrm{d}s \\ & - \phi_r'' \int_0^\ell \left(\chi_{12} + 2\chi_{10}\right) \mathrm{d}s - \left(\chi_{12}'' + 2\chi_{10}''\right) \int_0^\ell \phi_r \,\mathrm{d}s + \frac{3}{2k_0} \phi_r'' \int_0^\ell \phi_r'^2 \,\mathrm{d}s \Big]. \end{aligned} \tag{7.50}$$

By requiring $q_{21}(s)$ is orthogonal to $\phi_r(s)$, i.e., $\int_0^\ell \phi_r(s) q_{21}(s) \mathrm{d}s = 0$, and solving for $\partial_2 A_r$, the bifurcation equation is derived. When the quantities are back rescaled, the equation reads:

$$\dot{A}_r = i c_3 A_r^2 \bar{A}_r, \tag{7.51}$$

where, after integration by parts and accounting for boundary conditions:

$$\begin{aligned} c_3 := & \frac{EAk_0}{2m\ell\omega_r \int_0^\ell \phi_r^2 \,\mathrm{d}s} \left[-2 \int_0^\ell \phi_r \,\mathrm{d}s \int_0^\ell \phi_r' \left(\chi_{12}' + 2\chi_{10}' \right) \mathrm{d}s \right. \\ & \left. - \int_0^\ell \phi_r'^2 \,\mathrm{d}s \int_0^\ell (\chi_{12} + 2\chi_{10}) \,\mathrm{d}s + \frac{3}{2k_0} \left(\int_0^\ell \phi_r'^2 \,\mathrm{d}s \right)^2 \right]. \end{aligned} \tag{7.52}$$

(e) By letting $A_r = \frac{1}{2} a_r e^{i\varphi_r}$, and separating real and imaginary parts, two real equations are obtained:

$$\begin{aligned} \dot{a}_r &= 0, \\ a_r \dot{\varphi}_r &= \frac{1}{4} c_3 a_r^3, \end{aligned} \tag{7.53}$$

whose solution is: $a_r = \text{const}$, $\varphi_r = \Delta\omega_r t + \varphi_{r0}$, where $\Delta\omega_r := \frac{1}{4} c_3 a_r^2$. Since, at the leading order, is $v \simeq v_0 = a_r \phi_r(s) \cos(\omega_r t + \varphi_r(t)) = a_r \phi_r(s) \cos[(\omega_r + \Delta\omega_r) t + \varphi_{r0}]$, $\Delta\omega_r$ assumes the meaning of *frequency correction*. Therefore, $\varpi_r := \omega_r + \frac{1}{4} c_3 a_r^2$ is the amplitude-dependent *nonlinear frequency*. Table 7.1 reports the values of c_3 for the first symmetric/antisymmetric mode when $EA = 2.97 \times 10^7$ N, $\ell = 267$ m, $m = 1.8$ $\frac{\text{kg}}{\text{m}}$ and for different values of the Irvine parameter. In the symmetric mode, it is observed that for low values of the Irvine parameter λ^2, the cubic nonlinearity is dominant, resulting in an increase of the frequency (i.e., $\Delta\omega_r > 0$, because $c_3 > 0$). However, for larger values of

Table 7.1 Coefficient c_3 in the frequency-amplitude relationship of a suspended cable, for the first symmetric/antisymmetric mode, and for different values of the Irvine parameter; sag d in meters, with lenght $\ell = 267$ m fixed

d	$\frac{\lambda^2}{4\pi^2}$	1st mode	ω_1	c_3
1.807	0.0253	sym	2.692	0.027
2.606	0.0760	sym	2.405	0.0009
3.893	0.2533	sym	2.372	−0.058
5.615	0.7599	sym	2.681	−0.110
6.180	1.0132	antisym	2.799	0.407
6.428	1.1399	antisym	2.744	0.419
6.657	1.2665	antisym	2.697	0.434
7.448	1.7731	antisym	2.550	0.518

λ^2, the quadratic nonlinearity becomes dominant and leads to a decrease of the frequency (i.e., $\Delta\omega_r < 0$, because $c_3 < 0$). On the other hand, in the asymmetric mode, the cubic nonlinearity is dominant for all values of the Irvine parameter (i.e., $c_3 > 0$).

(f) The previous results can be particularized to a taut string, by letting the self-weight $mg = 0$. This entails that the initial curvature $k_0 = 0$ and the tension N_0, being an undermined $0/0$ ratio, is unrelated to the weight, but it is a free datum of the problem. Hence, all the modes, symmetric and antisymmetric, are sinusoidal, i.e., $\phi_r = \sin\left(\frac{r\pi}{\ell}s\right)$, $r = 1, 2, \ldots$, associated with the frequencies $\omega_r = \frac{r\pi}{\ell}\sqrt{\frac{N_0}{m}}$. Due to the vanishing of the quadratic nonlinearities, it is $\chi_{12}(s) \equiv 0$, $\chi_{10}(s) \equiv 0$ (and therefore $v_1 \equiv 0$) in Eq. 7.45. Accordingly, c_3 simplifies into:

$$c_3 := \frac{3EA}{\sqrt{mN_0}}\left(\frac{r\pi}{2\ell}\right)^3. \tag{7.54}$$

□

Problem 7.5 (*Primary external resonance of a clamped-sliding beam under prescribed motion at a support*) Consider a clamped-sliding beam, executing undamped nonlinear forced vibrations induced by the motion $v(\ell, t) = \alpha\breve{v}_B \cos(\Omega t)$ of the right support (Fig. 7.3b). (a) Write the equation which governs the motion. (b) Apply the MSM to derive the perturbation equations. (c) Obtain the bifurcation equation for Ω close to the fundamental frequency ω_1.

(a) Since the beam is axially unrestrained, its motion is ruled by Eq. 5.18. By zeroing the damping, the prestress and the external forces, it reduces to:

$$m\ddot{v} + EI\left\{v'''' + \left[v'\left(v'v''\right)'\right]'\right\} + \frac{1}{2}m\left[v'\int_{\ell}^{s}\left(\int_{0}^{s}\left(v'^2\right)^{\cdot\cdot}\mathrm{d}s\right)\mathrm{d}s\right]' = 0, \tag{7.55}$$

with the boundary conditions:

$$v(0,t) = 0, \quad v'(0,t) = 0, \quad v(\ell,t) = \alpha\breve{v}_B\cos(\Omega t), \quad v'(\ell,t) = 0. \tag{7.56}$$

To make homogeneous the boundary conditions, the following change of variable is introduced:

$$v(s,t) = \alpha\chi_0(s)\cos(\Omega t) + w(s,t), \tag{7.57}$$

in which:

$$\chi_0(s) = \breve{v}_B\left[3\left(\frac{s}{\ell}\right)^2 - 2\left(\frac{s}{\ell}\right)^3\right] \tag{7.58}$$

is the static linear response of the beam to the prescribed displacement $\breve{v}_B$. Accordingly, Eq. 7.57 expresses the motion of the beam as the sum of the known quasi-steady response, proportional to $\chi_0(s)$, and the unknown dynamic component $w(s,t)$. In the new variable, the Eq. 7.55 reads

$$m\ddot{w} + EI\left\{w'''' + \left[w'\left(w'w''\right)'\right]'\right\} + \frac{1}{2}m\left[w'\int_{\ell}^{s}\left(\int_{0}^{s}\left(w'^2\right)^{\cdot\cdot}\,\mathrm{d}s\right)\mathrm{d}s\right]' \tag{7.59}$$
$$= \alpha m\Omega^2\chi_0(s)\cos(\Omega t) + O(\alpha w)$$

and the boundary conditions, Eqs. 7.56, become homogeneous, i.e., $w = w' = 0$ at $x = 0, \ell$.

(b) Since the nonlinearities are only cubic, the rescaling $w \to \epsilon^{1/2}w$, $\alpha \to \epsilon^{3/2}\hat{\alpha}$ is performed. By dividing the equation by $\epsilon^{1/2}$, the nonlinearities and the excitation are shifted to the ϵ order. Independent time scales $t_0 := t, t_1 := \epsilon t, \ldots$ are introduced, so that $\ddot{w} = \left(\partial_0^2 + 2\epsilon\partial_0\partial_1 + \cdots\right)w$. Moreover, to express the closeness of excitation to the fundamental frequency, $\Omega = \omega_r + \epsilon\hat{\sigma}_r$ is posed. Finally, the displacement is expanded as $w = w_0 + \epsilon w_1 + \cdots$. The following perturbation equations are drawn:

Order ϵ^0:

$$m\partial_0^2 w_0 + EI w_0'''' = 0. \tag{7.60}$$

Order ϵ^1:

$$m\partial_0^2 w_1 + EI w_1'''' = -2m\partial_0\partial_1 w_0 - EI\left[w_0'\left(w_0'w_0''\right)'\right]' - \frac{1}{2}m\left[w_0'\int_{\ell}^{s}\left(\int_{0}^{s}\partial_0^2\left(w_0'^2\right)\mathrm{d}s\right)\mathrm{d}s\right]'$$
$$+\frac{1}{2}\alpha m\Omega^2\chi_0(s)\left(e^{i\sigma_r t_1}e^{i\omega_r t_0} + \text{c.c.}\right), \tag{7.61}$$

with the boundary conditions $w_k = 0$, $w_k' = 0$, $k = 0, 1, \ldots$ at $x = 0, \ell$. As a normalization condition, it is imposed that the ω_r-harmonic component of the response $w\left(\frac{\ell}{2}, t\right)$ is equal to $a_r(t)$.

(c) The monomodal solution to the perturbation Eq. 7.60 is:

$$w_0 = A_r(t_1, \ldots)\phi_r(s)e^{i\omega_r t_0} + \text{c.c.}, \tag{7.62}$$

where:

$$\phi_r := b_r\left[\frac{\cosh(\beta_r\ell) - \cos(\beta_r\ell)}{\sin(\beta_r\ell) - \sinh(\beta_r\ell)}(\sin(\beta_r s) - \sinh(\beta_r s)) + \cos(\beta_r s) - \cosh(\beta_r s)\right] \tag{7.63}$$

is the normalized mode of frequency $\omega_r = \beta_r^2\sqrt{\frac{EI}{m}}$, with β_r one of the infinite roots of the characteristic equation $\cos(\beta_r\ell)\cosh(\beta_r\ell) = 1$ and b_r is the normalization coefficient. For the first of them, i.e., the fundamental one, we have $\beta_1\ell = 4.73$ and $b_1 = -0.6297$.

When Eq. 7.62 is substituted in the next order perturbation Eqs. 7.61, this latter becomes:

$$m\partial_0^2 w_1 + EI w_1'''' = q_{11}(s)e^{i\omega_r t_0} + \text{N.R.T.} + \text{c.c.}, \tag{7.64}$$

where:

$$q_{11}(s) := -2im\omega_r\phi_r\partial_1 A_r + A_r^2\bar{A}_r\left[2m\omega_r^2\left(\phi_r'\int_0^s\phi_r'(\hat{s})^2\,\mathrm{d}\hat{s} + \phi_r''\int_\ell^s\int_0^{\check{s}}\phi_r'(\hat{s})^2\,\mathrm{d}\hat{s}\,\mathrm{d}\check{s}\right)\right.$$
$$\left. -3EI\left[\phi_r'\left(\phi_r'\phi_r''\right)'\right]'\right] + \frac{1}{2}\alpha m\Omega^2\chi_0 e^{it_1\sigma_r}. \tag{7.65}$$

For compatibility, the ω_r-frequency terms must be orthogonal to ϕ_r, i.e., $\int_0^\ell \phi_r(s)q_{21}(s)\,\mathrm{d}s = 0$. From this condition, by coming back to the unrescaled quantities, the bifurcation equation is derived, i.e.:

$$\dot{A}_r = ic_3A_r^2\bar{A}_r + ic_0\alpha e^{i\sigma_r t}, \tag{7.66}$$

where:

$$c_3 := \frac{1}{2m\omega_r\int_0^\ell\phi_r^2\,\mathrm{d}s}\left[-2m\omega_r^2\left(\int_0^\ell\phi_r(s)\,\phi_r'(s)\int_0^s\phi_r'(\hat{s})^2\,\mathrm{d}\hat{s}\,\mathrm{d}s\right.\right.$$
$$\left.\left.+\int_0^\ell\phi_r(s)\,\phi_r''(s)\int_\ell^s\int_0^{\check{s}}\phi_r'(\hat{s})^2\,\mathrm{d}\hat{s}\,\mathrm{d}\check{s}\,\mathrm{d}s\right) + 3EI\int_0^\ell\phi_r\left[\phi_r'\left(\phi_r'\phi_r''\right)'\right]'\,\mathrm{d}s\right], \tag{7.67}$$
$$c_0 := -\frac{\Omega^2\int_0^\ell\chi_0\phi_r\,\mathrm{d}s}{4\omega_r\int_0^\ell\phi_r^2\,\mathrm{d}s}.$$

For the fundamental mode, i.e., ϕ_1, we have $c_3 = -\frac{305.825}{\ell^4}\sqrt{\frac{EI}{m}}$. Since $c_3 < 0$, and damping is zero, the frequency-amplitude law assumes the shape in Fig. 3.1d. □

Problem 7.6 (*Primary external resonance of a transversely loaded hinged square plate*) Consider a viscoelastic plate, occupying the domain $\Omega := \{(x, y) \mid 0 \le x \le \ell,\ 0 \le y \le \ell\}$, hinged at the boundary Γ, normally constrained in-plane, loaded at the midpoint by a pulsating transverse force $\alpha\check{P}\cos(\Omega t)$ (Fig. 7.3c). (a) Write the equation of motion of the plate in the regime of finite displacements, by accounting for both external and internal damping. (b) Derive the perturbation equations. (c) Evaluate the undamped natural frequencies ω_{nm}. (d) Determine the bifurcation equation, for the force in primary external resonance with the fundamental frequency $\omega_r := \omega_{11}$.

(a) The out-of-plane equations of motion follow from Eq. 5.28, by zeroing the prestress, namely:

$$\begin{aligned}
&m\ddot{w} + c_e\dot{w} + D\nabla^4 w + D_v\nabla^4\dot{w}\\
&-C\Big[u_{,x}\,w_{,xx} + v_{,y}\,w_{,yy} + \nu\,u_{,x}\,w_{,yy} + \nu\,v_{,y}\,w_{,xx} + (1-\nu)u_{,y}\,w_{,xy}\\
&+(1-\nu)v_{,x}\,w_{,xy}\Big] - C\Big[\frac{1}{2}w_{,x}^2\,w_{,xx} + \frac{1}{2}w_{,y}^2\,w_{,yy} + \frac{1}{2}\nu\,w_{,y}^2\,w_{,xx}\\
&+\frac{1}{2}\nu\,w_{,x}^2\,w_{,yy} + (1-\nu)\,w_{,x}\,w_{,y}\,w_{,xy}\Big] = \delta\left(x-\frac{\ell}{2}, y-\frac{\ell}{2}\right)\alpha\check{P}\cos(\Omega t).
\end{aligned} \tag{7.68}$$

where $\delta(\cdot)$ is the Dirac delta. The in-plane balance equations are given by Eqs. 7.8. The boundary conditions (see Problem 7.3) prescribe:

$$u = 0, \qquad \text{at } x = 0,\ \ell, \tag{7.69a}$$

$$v = 0, \qquad \text{at } y = 0,\ \ell, \tag{7.69b}$$

$$\tilde{N}_{xy} = C\check{\nu}\left(u,_y + v,_x + w,_x\, w,_y\right) = 0, \qquad \text{at } x,\ y = 0,\ \ell, \tag{7.69c}$$

$$w = 0, \qquad \text{at } x,\ y = 0,\ \ell, \tag{7.69d}$$

$$w_{,xx} = 0, \qquad \text{at } x = 0,\ \ell, \tag{7.69e}$$

$$w_{,yy} = 0, \qquad \text{at } y = 0,\ \ell. \tag{7.69f}$$

(b) The following rescaling is introduced: $(u, v, w) \rightarrow \epsilon\left(\hat{u}, \hat{v}, \hat{w}\right)$, $(c_e, D_v) \rightarrow \epsilon^2\left(\hat{c}_e, \hat{D}_v\right)$, $\alpha \rightarrow \epsilon^3\hat{\alpha}$, and the closeness between Ω and ω_r is expressed by $\Omega = \omega_r + \epsilon^2\hat{\sigma}_r$. Moreover, the time scales $t_0 := t$, $t_2 := \epsilon^2 t$, $\cdots$ are introduced, so that $\dot{w} = \left(\partial_0 + \epsilon^2\partial_2\right) w$ and $\ddot{w} = \left(\partial_0^2 + 2\partial_0\partial_2 + \cdots\right) w$. Finally, displacement are expanded as in Eq. 7.10. The following perturbation equations are derived:

Order ϵ^0:

$$m\partial_0^2 w_0 + D\nabla^4 w_0 = 0. \tag{7.70}$$

Order ϵ^1:

$$u_{1,xx} + \check{\nu} u_{1,yy} + \hat{\nu} v_{1,xy} = -w_{0,x} w_{0,xx} - \check{\nu} w_{0,x} w_{0,yy} - \hat{\nu} w_{0,y} w_{0,xy}, \tag{7.71a}$$

$$\hat{\nu} u_{1,xy} + v_{1,yy} + \check{\nu} v_{1,xx} = -\check{\nu} w_{0,y} w_{0,xx} - w_{0,y} w_{0,yy} - \hat{\nu} w_{0,x} w_{0,xy}. \tag{7.71b}$$

Order ϵ^2:

$$\begin{aligned}
m\partial_0^2 w_2 + D\nabla^4 w_2 =& -2m\partial_0\partial_2 w_0 - \left(c_e + D_v\nabla^4\right)\partial_0 w_0 \\
&+ C\Big[u_1,_x\, w_0,_{xx} + v_1,_y\, w_0,_{yy} + \nu\, u_1,_x\, w_0,_{yy} \\
&+ \nu\, v_1,_y\, w_0,_{xx} + (1-\nu) u_1,_y\, w_0,_{xy} + (1-\nu) v_1,_x\, w_0,_{xy}\Big] \\
&+ C\Big[\frac{1}{2} w_0,_x^2\, w_0,_{xx} + \frac{1}{2} w_0,_y^2\, w_0,_{yy} + \frac{1}{2}\nu\, w_0,_y^2\, w_0,_{xx} \\
&+ \frac{1}{2}\nu\, w_0,_x^2\, w_0,_{yy} + (1-\nu)\, w_0,_x\, w_0,_y\, w_0,_{xy}\Big] \\
&+ \frac{1}{2}\delta\left(x - \frac{\ell}{2}, y - \frac{\ell}{2}\right)\alpha\check{P}\left(e^{i\sigma_r t_2} e^{i\omega_r t_0} + \text{c.c.}\right),
\end{aligned} \tag{7.72}$$

together with the boundary conditions on Γ:

$$w_k = 0, \quad \text{at } x, y = 0, \ \ell, \tag{7.73a}$$

$$w_{k,xx} = 0, \quad \text{at } x = 0, \ \ell, \qquad k = 0, 2, \tag{7.73b}$$

$$w_{k,yy} = 0, \quad \text{at } y = 0, \ \ell, \tag{7.73c}$$

and

$$u_1 = 0, \quad \text{at } x = 0, \ \ell, \tag{7.74a}$$

$$v_1 = 0, \quad \text{at } y = 0, \ \ell, \tag{7.74b}$$

$$C\check{\nu}\left(u_{1,y} + v_{1,x}\right) = -C\check{\nu} w_{0,x}\, w_{0,y}, \quad \text{at } x, y = 0, \ \ell. \tag{7.74c}$$

As a normalization condition, it is required that the ω_r-harmonic component of the response $w\left(\frac{\ell}{2}, \frac{\ell}{2}, t\right)$ is equal to $a_r(t)$.

(c) The ϵ^0 order perturbation Eq. 7.70, with relevant boundary conditions, admits the following monomodal solution[7]:

$$w_0 = A_{hk}\left(t_2, \ldots\right) \phi_{hk}\left(x, y\right) e^{i\omega_{hk} t_0} + \text{c.c.}, \tag{7.75}$$

where $\omega_{hk} = \left(h^2 + k^2\right) \frac{\pi^2}{\ell^2} \sqrt{\frac{D}{m}}$ and $\phi_{hk} = \sin\left(\frac{h\pi x}{\ell}\right) \sin\left(\frac{k\pi y}{\ell}\right)$. The fundamental frequency $(h = k = 1)$ is $\omega_r := \omega_{11} = 2\frac{\pi^2}{\ell^2}\sqrt{\frac{D}{m}}$ and the relevant mode:

$$\phi_r = \sin\left(\frac{\pi x}{\ell}\right) \sin\left(\frac{\pi y}{\ell}\right); \tag{7.76}$$

the corresponding amplitude is called A_r. By substituting w_0 in the perturbation Eqs. 7.71, these latter read (remember the definitions $\hat{\nu} := \frac{1+\nu}{2}$ and $\check{\nu} := \frac{1-\nu}{2}$):

$$u_{1,xx} + \check{\nu} u_{1,yy} + \hat{\nu} v_{1,xy} = -\frac{\pi^3}{2\ell^3}\left[\cos\left(\frac{2\pi y}{\ell}\right) - \check{\nu}\right] \sin\left(\frac{2\pi x}{\ell}\right)\left(A_r^2 e^{2i\omega_r t_0} + A_r \bar{A}_r\right) + \text{c.c.}, \tag{7.77a}$$

$$\hat{\nu} u_{1,xy} + v_{1,yy} + \check{\nu} v_{1,xx} = -\frac{\pi^3}{2\ell^3}\left[\cos\left(\frac{2\pi x}{\ell}\right) - \check{\nu}\right] \sin\left(\frac{2\pi y}{\ell}\right)\left(A_r^2 e^{2i\omega_r t_0} + A_r \bar{A}_r\right) + \text{c.c.}, \tag{7.77b}$$

whose solution is:

$$u_1 = U_1\left(x, y\right)\left(A_r^2 e^{2i\omega_r t_0} + A_r \bar{A}_r\right) + \text{c.c.}, \tag{7.78a}$$

$$v_1 = V_1\left(x, y\right)\left(A_r^2 e^{2i\omega_r t_0} + A_r \bar{A}_r\right) + \text{c.c.}, \tag{7.78b}$$

where:

[7] Here the usual integer numebers n, m have been changed into h, k, to avoid confusion with the mass density.

$$U_1(x,y) := \frac{\pi}{16\ell}\left[\cos\left(\frac{2\pi y}{\ell}\right) - 2\check{\nu}\right]\sin\left(\frac{2\pi x}{\ell}\right), \tag{7.79a}$$

$$V_1(x,y) := \frac{\pi}{16\ell}\left[\cos\left(\frac{2\pi x}{\ell}\right) - 2\check{\nu}\right]\sin\left(\frac{2\pi y}{\ell}\right), \tag{7.79b}$$

similar to Eqs. 7.17b, c. With the the previous results, the perturbation Eq. 7.72 becomes:

$$m\partial_0^2 w_2 + D\nabla^4 w_2 = q_{11}(x,y)\, e^{i\omega_r t_0} + \text{N.R.T.} + \text{c.c.}, \tag{7.80}$$

where:

$$q_{11}(x,y) := -2i\omega_r m\partial_2 A_r\phi_r - i\omega_r A_r\left(c_e + \frac{4\pi^4}{\ell^4}D_v\right)\phi_r - 3\frac{C\pi^4\hat{\nu}}{2\ell^4}A_r^2\bar{A}_r\left[1 - \check{\nu}\cos\left(\frac{2\pi x}{\ell}\right)\right.$$
$$\left. -\check{\nu}\cos\left(\frac{2\pi y}{\ell}\right)\right]\phi_r + \frac{1}{2}\delta\left(x - \frac{\ell}{2}, y - \frac{\ell}{2}\right)\alpha\check{P}e^{i\sigma_r t_2}. \tag{7.81}$$

By requiring the resonant terms are orthogonal to ϕ_r, i.e., $\int_0^\ell\int_0^\ell \phi_r(x,y)\, q_{11}(x,y)\, \mathrm{d}y\, \mathrm{d}x = 0$, and coming back to the unrescaled quantities, the bifurcation equation follows:

$$\dot{A}_r = -c_1 A_r + ic_3 A_r^2\bar{A}_r + ic_0\alpha e^{i\sigma_r t}, \tag{7.82}$$

where:

$$c_1 := \frac{1}{2m}\left(c_e + \frac{4\pi^4}{\ell^4}D_v\right), \quad c_3 := \frac{3}{4}\frac{\pi^4}{\ell^4}\frac{C}{m\omega_r}\hat{\nu}\left(1+\check{\nu}\right), \quad c_0 := -\frac{\check{P}}{m\omega_r\ell^2}. \tag{7.83}$$

Since $c_3 > 0$, the frequency-amplitude law assumes the shape represented in Fig 3.1a. □

Problem 7.7 (*Sub- and super-harmonic resonances of a clamped-clamped beam*) A viscoelastic clamped-clamped beam, is loaded by a transverse uniformly distributed load $\alpha\check{p}_y\cos(\Omega t)$ (Fig. 7.3d). (a) Write the equation of motion of the beam in the finite displacement regime. (b) By applying the MSM, obtain the perturbation equations, by ordering the external load at the leading order and shifting damping at the same order of nonlinearities. (c) Solve the bifurcation equation, and detect all the possible resonance cases. (d) Derive the bifurcation equation when: (i) the excitation frequency is in 3:1 sub-harmonic resonance with the rth frequency ω_r, i.e., when $\Omega \simeq 3\omega_r$; (ii) when the excitation frequency is in 1:3 super-harmonic resonance with ω_r, i.e., when $\Omega \simeq \frac{1}{3}\omega_r$.

(a) Since the beam is axially restrained, the motion is ruled by Eq. 5.11, whit the prestress $N_0 = 0$, i.e.:

$$m\ddot{v} + c_e\dot{v} + EIv'''' + \eta I\dot{v}'''' - \frac{EA}{2\ell}v''\int_0^\ell v'^2\mathrm{d}s = \alpha\check{p}_y\cos(\Omega t), \tag{7.84}$$

with the boundary conditions $v = 0$, $v' = 0$ at $x = 0, \ell$.

(b) By performing the rescaling $v \to \epsilon^{1/2}\hat{v}$ and dividing by $\epsilon^{1/2}$, the cubic nonlinearities are shifted to the ϵ order; moreover, via the rescaling $(c_e, \eta) \to \epsilon\left(\hat{c}_e, \hat{\eta}\right)$, damping is ordered at the same order of cubic nonlinearities. In contrast, since the excitation is *not* primarily resonant, it is considered of the same order of displacements, i.e., $\alpha \to \epsilon^{1/2}\hat{\alpha}$. Independent time scales $t_0 := t$, $t_1 := \epsilon t$, $\cdots$ are then introduced, so that $\dot{v} = \left(\partial_0 + \epsilon^2\partial_1\right) v$ and $\ddot{v} = \left(\partial_0^2 + 2\epsilon\partial_0\partial_1 + \cdots\right) v$. Finally, the displacement is expanded as $v = v_0 + \epsilon v_1 + \cdots$. As a normalization condition, it is required that the ω_r-harmonic component of the response $v\left(\frac{\ell}{2}, t\right)$ is equal to $a_r(t)$. The following perturbation equations are drawn:

Order ϵ^0:

$$m\partial_0^2 v_0 + EI v_0'''' = \frac{1}{2}\alpha\breve{p}_y\left(e^{i\Omega t_0} + \text{c.c.}\right). \tag{7.85}$$

Order ϵ^1:

$$m\partial_0^2 v_1 + EI v_1'''' = -2m\partial_0\partial_1 v_0 - c_e\partial_0 v_0 - \eta I\partial_0 v_0'''' + \frac{EA}{2\ell} v_0'' \int_0^\ell v_0'^2 \mathrm{d}s, \tag{7.86}$$

with the boundary conditions $v_k = 0$, $v_k' = 0$ $k = 0, 1, \ldots$ at $x = 0, \ell$.

(c) The solution to the generating perturbation Eq. 7.85 reads:

$$v_0 = A_r\left(t_1, \ldots\right)\phi_r(s)\, e^{i\omega_r t_0} + \frac{1}{2}\alpha\chi_0(s)\, e^{i\Omega t_0} + \text{c.c.}, \tag{7.87}$$

where ϕ_r, ω_r have already been introduced in Eqs. 7.62, 7.63, and, moreover with:

$$\chi_0 := \frac{\breve{p}_y}{m\Omega^2}\left[-1 + b_1\cos(\beta s) + b_2\sin(\beta s) + b_3\cosh(\beta s) + b_4\sinh(\beta s)\right], \tag{7.88}$$

where $\beta := \sqrt[4]{\frac{m\Omega^2}{EI}}$ and:

$$\begin{aligned} b_1 &:= \frac{\tanh\left(\frac{\beta\ell}{2}\right)}{\tan\left(\frac{\beta\ell}{2}\right) + \tanh\left(\frac{\beta\ell}{2}\right)}, \quad b_2 := \frac{\tan\left(\frac{\beta\ell}{2}\right)\tanh\left(\frac{\beta\ell}{2}\right)}{\tan\left(\frac{\beta\ell}{2}\right) + \tanh\left(\frac{\beta\ell}{2}\right)}, \\ b_3 &:= \frac{\tan\left(\frac{\beta\ell}{2}\right)}{\tan\left(\frac{\beta\ell}{2}\right) + \tanh\left(\frac{\beta\ell}{2}\right)}, \quad b_4 := -b_2. \end{aligned} \tag{7.89}$$

With Eqs. 7.86, 7.87 transform into:

$$m\ddot{v}_1 + EI v_1'''' = q_{11}(s)\, e^{i\omega_r t_0} + q_{12}(s)\, e^{3i\Omega t_0} + q_{13}(s)\, e^{i(\Omega - 2\omega_r)t_0} + \text{c.c.} + \text{N.R.T.}, \tag{7.90}$$

in which:

$$q_{11} := -2im\omega_r\phi_r\partial_1 A_r - ic_e\omega_r A_r\phi_r - i\eta I\omega_r A_r\phi_r''''$$
$$+\frac{3EA}{2\ell}A_r^2\bar{A}_r\phi_r''\int_0^\ell \phi_r'^2 \mathrm{d}s + \alpha^2\frac{EA}{4\ell}A_r\left(\phi_r''\int_0^\ell \chi_0'^2\,\mathrm{d}s + 2\chi_0''\int_0^\ell \chi_0'\phi_r'\,\mathrm{d}s\right),$$
$$q_{12} := \alpha^3\frac{EA}{16\ell}\chi_0''\int_0^\ell \chi_0'^2\,\mathrm{d}s,$$
$$q_{13} := \alpha\frac{EA}{4\ell}\bar{A}_r^2\left(\chi_0''\int_0^\ell \phi_r'^2\,\mathrm{d}s + 2\phi_r''\int_0^\ell \chi_0'\phi_r'\,\mathrm{d}s\right). \tag{7.91}$$

Due to the hypothesis $\Omega \neq \omega_r$, the frequencies $\Omega, 3\omega_r, \Omega + 2\omega_r, 2\Omega \pm \omega_r$, that appear in the right side of Eq. 7.90, are surely different from ω_r, and therefore classified as non-resonant. Among the remaining excitation frequencies involving Ω, two cases are possible:

(i) $\Omega - 2\omega_r = \omega_r$, when $\Omega = 3\omega_r$ (sub-harmonic resonance, since the natural frequency is smaller than the excitation frequency);

(ii) $3\Omega = \omega_r$, when $\Omega = \frac{1}{3}\omega_r$ (super-harmonic resonance, since the natural frequency is larger than the excitation frequency).

(d) To analyze the *sub-harmonic* resonance, $\Omega = 3\omega_r + \epsilon\hat{\sigma}_r$ is introduced in the q_{13}-term of Eq. 7.90 and the following compatibility condition enforced:

$$\int_0^\ell \phi_r(s)\left[q_{11}(s) + q_{13}(s)\,e^{i\sigma_r t_1}\right]\mathrm{d}s = 0, \tag{7.92}$$

from which the bifurcation equation is drawn:

$$\dot{A}_r = -c_1 A_r + ic_3 A_r^2\bar{A}_r + ic_{21}\alpha^2 A_r + ic_{12}\alpha\bar{A}_r^2 e^{i\sigma_r t}, \tag{7.93}$$

where:

$$c_1 := \frac{1}{2m}\left(c_e + \eta I\frac{\int_0^\ell \phi_r\phi_r''''\,\mathrm{d}s}{\int_0^\ell \phi_r^2\,\mathrm{d}s}\right),$$
$$c_3 := -\frac{3EA}{4m\ell\omega_r}\frac{\int_0^\ell \phi_r'^2\,\mathrm{d}s\int_0^\ell \phi_r\phi_r''\,\mathrm{d}s}{\int_0^\ell \phi_r^2\,\mathrm{d}s},$$
$$c_{21} := -\frac{EA}{8m\ell\omega_r\int_0^\ell \phi_r^2\,\mathrm{d}s}\left(\int_0^\ell \phi_r\phi_r''\,\mathrm{d}s\int_0^\ell \chi_0'^2\,\mathrm{d}s + 2\int_0^\ell \phi_r\chi_0''\,\mathrm{d}s\int_0^\ell \phi_r'\chi_0'\,\mathrm{d}s\right),$$
$$c_{12} := -\frac{EA}{8m\ell\omega_r\int_0^\ell \phi_r^2\,\mathrm{d}s}\left(\int_0^\ell \phi_r\chi_0''\,\mathrm{d}s\int_0^\ell \phi_r'^2\,\mathrm{d}s + 2\int_0^\ell \phi_r\phi_r''\,\mathrm{d}s\int_0^\ell \phi_r'\chi_0'\,\mathrm{d}s\right). \tag{7.94}$$

To analyze the *super-harmonic* resonance, $3\Omega = \omega_r + \epsilon\hat{\sigma}_r$ is introduced in the q_{12}-term of Eq. 7.90 and the following compatibility condition enforced:

$$\int_0^\ell \phi_1(s)\left[q_{11}(s) + q_{12}(s)\, e^{i\sigma_r t_1}\right] \mathrm{d}s = 0, \tag{7.95}$$

from which the bifurcation equation is drawn:

$$\dot{A}_r = -c_1 A_r + i c_3 A_r^2 \bar{A}_r + i c_{21}\alpha^2 A_r + i c_{30}\alpha^3 e^{i\sigma_r t}, \tag{7.96}$$

where c_1, c_3 and c_{21} are still given by Eqs. 7.94, while:

$$c_{30} := -\frac{EA}{32 m \ell \omega_r \int_0^\ell \phi_r^2 \,\mathrm{d}s} \int_0^\ell \phi_r \chi_0'' \,\mathrm{d}s \int_0^\ell \chi_0'^2 \,\mathrm{d}s. \tag{7.97}$$

□

Problem 7.8 (*Nonlinear behavior of axially unrestrained versus axially restrained beams*) To analyze the nonlinear effects induced by the axial constraints, consider a clamped-sliding (Fig. 7.3b) and a clamped-clamped beam (Fig. 7.3d), i.e., beams equally constrained against transverse displacements, but not against longitudinal displacements. By using the results achieved in Problems 7.5 and 7.7: (a) write the free vibration amplitude-frequency relationship for both cases. (b) Comment the dependence on mechanical parameters of the two relationships. (c) Plot and compare the nonlinear frequency versus the amplitude, to state the entity of the nonlinear effects.

(a) The bifurcation equation for undamped free vibrations of nonlinear systems is:

$$\dot{A}_r = i c_3 A_r^2 \bar{A}_r, \tag{7.98}$$

as already found in Eq. 7.51 for a cable, in Eq. 7.66 for a clamped-sliding beam (by letting $\alpha = 0$), and in Eqs. 7.93, 7.96 for a clamped-clamped beam (by vanishing damping and letting $\alpha = 0$). Since $A_r = \frac{1}{2} a_r e^{i\varphi_r}$, Eq. 7.98 provides two real equations, namely (Eq. 7.53), $\dot{a}_r = 0$ and $a_r \dot{\varphi}_r = \frac{1}{4} c_3 a_r^3$, from which the amplitude-frequency relationship follows:

$$\varpi_r := \omega_r + \frac{1}{4} c_3 a_r^2. \tag{7.99}$$

Here ϖ_r is the nonlinear counterpart of the linear frequency $\omega_r = \beta_r^2 \sqrt{\frac{EI}{m}}$, with β_r one of the infinite roots of the characteristic equation $\cos(\beta_r \ell)\cosh(\beta_r \ell) = 1$. For the two types of beams, it has been found (Eqs. 7.67a and 7.94b, respectively), that:

$$c_3 := \begin{cases} \dfrac{1}{2m\omega_r \int_0^\ell \phi_r^2 \, \mathrm{d}s} \Bigg[-2m\omega_r^2 \Bigg(\displaystyle\int_0^\ell \phi_r(s)\, \phi_r'(s) \int_0^s \phi_r'(\hat{s})^2 \, \mathrm{d}\hat{s}\, \mathrm{d}s \\ + \displaystyle\int_0^\ell \phi_r(s)\, \phi_r''(s) \int_\ell^s \int_0^{\check{s}} \phi_r'(\hat{s})^2 \, \mathrm{d}\hat{s}\, \mathrm{d}\check{s}\, \mathrm{d}s \Bigg) & \text{clamped} - \text{sliding}, \\ + 3EI \displaystyle\int_0^\ell \phi_r \left[\phi_r' \left(\phi_r' \phi_r''\right)' \right]' \mathrm{d}s \Bigg], \\ -\dfrac{3EA}{4m\ell\omega_r} \dfrac{\int_0^\ell \phi_r'^2 \, \mathrm{d}s \int_0^\ell \phi_r \phi_r'' \, \mathrm{d}s}{\int_0^\ell \phi_r^2 \, \mathrm{d}s}, & \text{clamped} - \text{clamped}. \end{cases} \tag{7.100}$$

which for the foundamental modes ϕ_1 becomes:

$$c_3 := \begin{cases} -\dfrac{305.825}{\ell^4}\sqrt{\dfrac{EI}{m}}, & \text{clamped} - \text{sliding}, \\ 2.011 \dfrac{EA}{EI\ell^2}\sqrt{\dfrac{EI}{m}}, & \text{clamped} - \text{clamped}. \end{cases} \tag{7.101}$$

The magnitude of nonlinearity in Eq. 7.99 is encompassed by the c_3 coefficient: the larger it is in modulus, the larger the nonlinear effect.

(b) To discuss the frequency-amplitude relationship, it is convenient to recast it in nondimensional form, i.e.:

$$\frac{\varpi_r}{\omega_r} = 1 + k\left(\frac{a_r}{L}\right)^2, \qquad k := \frac{c_3 L^2}{4\omega_r}, \tag{7.102}$$

where L is a length, to be properly selected. Based on the previous expression, and according to the relevant equations of motion, the following remarks hold.

In the sliding beam, two contribution can be noticed, linked to different mechanisms: (i) the elastic contribution, which is induced by the nonlinear curvature, producing a *hardening* behavior; (ii) the inertial contribution, which is related to the longitudinal motion, triggering a *softening* behavior. The resulting nonlinear effect is the algebraic sum of the two antagonist contributions. Since the nonlinear frequency only depends on the flexural stiffness EI, it is convenient to take $L = \ell$ in Eq. 7.102, in order $\frac{\varpi_r}{\omega_r}$ becomes independent of any mechanical parameters.[8] Hence:

$$\frac{\varpi_r}{\omega_r} = 1 + k\left(\frac{a_r}{\ell}\right)^2, \qquad k := -3.4173. \tag{7.103}$$

In the fixed beam only one contribution is noticed, of elastic nature, since the longitudinal inertia has been neglected in the model; namely, the stretching effect (triggered by bending), which induces a *hardening* behavior. Since the frequency depends, in addition to EI, also on the $\lambda := \sqrt{\frac{EA\ell^2}{EI}}$ *slenderness ratio* it is convenient to take $L = \varrho$ in Eq. 7.102, with $\varrho = \sqrt{\frac{I}{A}}$ the inertia radius, in order $\frac{\varpi_r}{\omega_r}$ becomes independent of any mechanical parameters. Hence:

[8] Remember that ω_r is proportional to $\sqrt{\frac{EI}{m}}$.

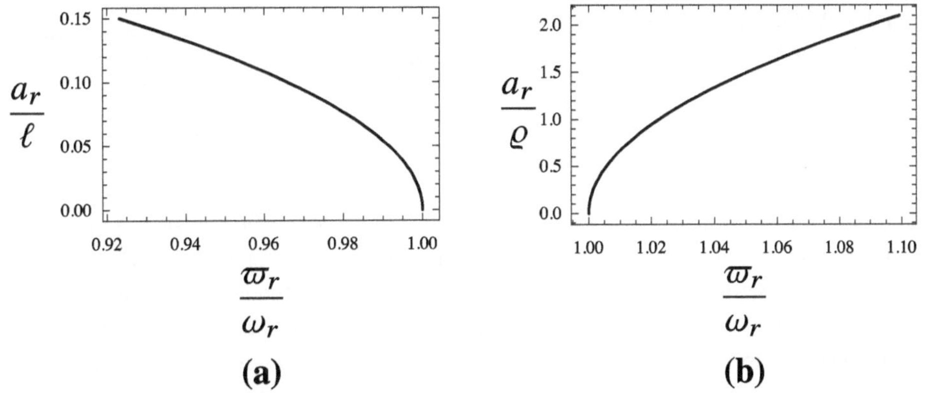

Fig. 7.4 Nonlinear frequency-amplitude relationship for: **a** clamped-sliding beam, **b** clamped-clamped beam. Amplitudes nondimensionalized with respect the length and the inertia radius, respectively

$$\frac{\varpi_r}{\omega_r} = 1 + k\left(\frac{a_r}{\varrho}\right)^2, \qquad k := 0.0224733. \tag{7.104}$$

(c) By using Eq. 7.103, the nondimensional nonlinear frequency $\frac{\varpi_r}{\omega_r}$ of the sliding beam is plotted in Fig. 7.4a, *versus* the amplitude-to-length ratio. It is seen that large amplitudes (e.g., 15% of the length) are needed to appreciate a frequency correction of few percent units. This is due to the intrinsic low nonlinearity of the system, and to the occurrence of the two antagonist mechanisms cited. Since the inertial effects prevail, the beam exhibit a softening behavior.[9]

By using Eq. 7.104, the frequency-amplitude relationship for the fixed beam is plotted in Fig. 7.4b. It appears that: (i) the behavior, as expected, is hardening; (ii) the nonlinearities strongly affect the mechanical behavior, resulting in a frequency correction of about 10% with an amplitude-to-radius ratio of about 2 (i.e., taken for example $\frac{\ell}{\varrho} = 100$, with an amplitude $a_r = \frac{\ell}{50}$). □

7.5 Parametric Excitation

As an example of parametric excitation, the Bolotin beam in the principal resonance condition is studied, for which the instability region is determined and the amplitude of the limit cycle computed (Fig. 7.5).

[9] Smaller, in absolute value, nonlinear corrections are obtained for different boundary conditions, but even nearly zero correction for the clamped-free beam, which exhibits also a behavior of hardening type, as shown in [3].

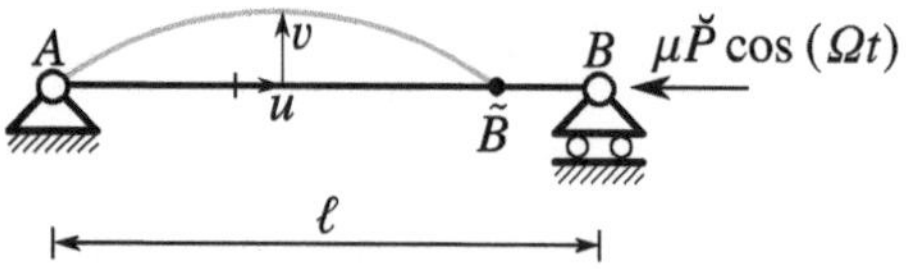

Fig. 7.5 Parametric excitation of the Bolotin beam

Problem 7.9 (*Parametric excitation of the damped Bolotin beam*) Consider a viscoelastic hinged-rolled beam, compressed by a harmonically varying force $P = \mu\breve{P}\cos(\Omega t)$, in principal parametric resonance condition, $\Omega \simeq 2\omega_r$, with the fundamental frequency $\omega_r := \omega_1$ (Fig. 7.5). (a) Write the equation of motion. (b) Apply the MSM to derive the perturbation equations. (c) Obtain the bifurcation equation and, by exploiting the results of the Chap. 4, determine the region of instability and the amplitude of the limit cycle.

(a) Since the beam is axially unrestrained, its equation of motion is provided by Eq. 5.18, with the geometric term $Pv''\left(1+\frac{3}{2}v'^2\right)$ depending on time, and the incremental external force vanished, namely:

$$m\ddot{v} + c_e\dot{v} + EI\left\{v'''' + \left[v'\left(v'v''\right)'\right]'\right\} + \eta I\dot{v}'''' + \mu\breve{P}\cos(\Omega t)\,v''\left(1+\frac{3}{2}v'^2\right)$$
$$+\frac{1}{2}m\left[v'\int_\ell^s\left(\int_0^s \left(v'^2\right)^{\cdot\cdot}\,\mathrm{d}s\right)\mathrm{d}s\right]' = 0, \tag{7.105}$$

to be integrated with the boundary conditions $v(0,t) = v(\ell,t) = 0$ and $EIv''(0,t)(1+\frac{1}{2}v'^2(0,t)) + \eta I\dot{v}''(0,t) = 0$, $EIv''(\ell,t)(1+\frac{1}{2}v'^2(\ell,t)) + \eta I\dot{v}''(\ell,t) = 0$.

(b) To apply the MSM, the rescaling $v \to \epsilon^{1/2}\hat{v}$, $\mu \to \epsilon\hat{\mu}$, $(c_e,\eta) \to \epsilon\left(\hat{c}_e,\hat{\eta}\right)$ is performed. Moreover, the time-scales $t_0 := t$, $t_1 := \epsilon t$ are introduced, for which $\dot{v} = (\partial_0 + \epsilon\partial_1 + \cdots)\,v$, $\ddot{v} = \left(\partial_0^2 + 2\epsilon\partial_0\partial_1 + \cdots\right)v$. Furthermore, the parametric frequency is expressed as $\Omega = 2\omega_r + \epsilon\hat{\sigma}_r$. Finally, the displacement is expanded as: $v = v_0 + \epsilon\, v_1 + \cdots$. The following perturbation equations are derived:

Order ϵ^0:

$$m\partial_0^2 v_0 + EIv_0'''' = 0. \tag{7.106}$$

Order ϵ^1:

$$m\partial_0^2 v_1 + EIv_1'''' = -2m\partial_0\partial_1 v_0 - c_e\partial_0 v_0 - EI\left[v_0'\left(v_0'v_0''\right)'\right]' - \eta I\partial_0 v_0''''$$
$$-\frac{1}{2}\mu\breve{P}\left(e^{i\sigma_r t_1}e^{2i\omega_r t_0} + \text{c.c.}\right)v_0'' - \frac{1}{2}m\left[v_0'\int_\ell^s\left(\int_0^s \partial_0^2\left(v_0'^2\right)\mathrm{d}s\right)\mathrm{d}s\right]', \tag{7.107}$$

with the geometric boundary conditions $v_k(0,t_k) = v_k(\ell,t_k) = 0$, $k = 0, 1, \ldots$, and the mechanical boundary conditions $EIv_0''(0,t_k) = EIv_0''(\ell,t_k) = 0$, $EIv_1''(0,t_k) + \eta I\partial_0 v_0''(0,t_k) +$

$\cancel{\frac{1}{2}EI v_0''(0,t_k)v_0'^2(0,t_k)} = 0$, $EI v_1''(\ell, t_k) + \cancel{\eta I \partial_0 v_0''(\ell, t_k)} + \cancel{\frac{1}{2}EI v_0''(\ell, t_k)v_0'^2(\ell, t_k)} = 0$ (in which the barred terms vanish because of the lower order condition). Concerning normalization, it is required that the ω_r-harmonic component of the response $v\left(\frac{\ell}{2}, t\right)$ is equal to $a_r(t)$.

(c) The solution to Eq. 7.106, by accounting only for the fundamental mode ϕ_r $(r = 1)$ involved in the parametric resonance (the remaining being damped), reads:

$$v_0 = A_r(t_1, \ldots)\, \phi_r(s)\, e^{i\omega_r t_0} + \text{c.c.}, \tag{7.108}$$

where $\phi_r = \sin\left(\frac{\pi s}{\ell}\right)$ and $\omega_r = \frac{\pi^2}{\ell^2}\sqrt{\frac{EI}{m}}$. Consequently, Eqs. 7.107 become:

$$m\partial_0^2 v_1 + EI v_1'''' = q_{11}(s)\, e^{i\omega_r t_0} + \text{N.R.T.} + \text{c.c.}, \tag{7.109}$$

where:

$$\begin{aligned} q_{11} := -\Bigg\{ & 2i\omega_r m \partial_1 A_r \phi_r + i c_e \omega_r A_r \phi_r + i\,\eta I \omega_r A_r \phi_r'''' + \frac{1}{2}\mu \breve{P} \bar{A}_r e^{i\sigma_r t_1} \phi_r'' \\ & - A_r^2 \bar{A}_r \Bigg[2m\omega_r^2 \left(\phi_r' \int_0^s \phi_r'(\hat{s})^2\, \mathrm{d}\hat{s} + \phi_r'' \int_\ell^s \int_0^{\check{s}} \phi_r'(\hat{s})^2\, \mathrm{d}\hat{s}\, \mathrm{d}\check{s} \right) \\ & - 3EI \left[\phi_r' \left(\phi_r' \phi_r'' \right)' \right]' \Bigg] \Bigg\}. \end{aligned} \tag{7.110}$$

Solvability requires that $\int_0^\ell \phi_r(s)\, q_{11}(s)\, \mathrm{d}s = 0$, from which $\partial_1 A_r$ is drawn. By performing a backward rescaling, the bifurcation equation (formally identical to Eq. 3.71 after the replacements $c_1\xi \to c_1$ and $\nu \to \mu$) is finally obtained:

$$\dot{A}_r = -c_1 A_r + i c_3 A_r^2 \bar{A}_r + i c_0 \mu \bar{A}_r e^{i\sigma_r t}, \tag{7.111}$$

where the coefficients are defined as follows:

$$c_0 := -\frac{\breve{P}}{4\sqrt{mEI}}, \quad c_1 := \frac{1}{2m}\left(c_e + \frac{\pi^4}{\ell^4}\eta I\right), \quad c_3 := -\frac{\pi^4}{\ell^4}\left(\frac{\pi^2}{6} - \frac{15}{16}\right)\sqrt{\frac{EI}{m}}. \tag{7.112}$$

The analysis proceeds as in Chap. 3. In particular, by exploiting the results of Eq. 3.83, again with substitutions $c_1\xi \to c_1$ and $\nu \to \mu$, the primary instability zone is found to be bounded by the curve:

$$\mu = \frac{4\sqrt{mEI}}{\breve{P}}\sqrt{\frac{1}{4m^2}\left(c_e + \frac{\pi^4}{\ell^4}\eta I\right)^2 + \frac{\sigma_r^2}{4}}. \tag{7.113}$$

Moreover, by using Eq. 3.87, the limit cycle amplitude is:

$$a_{re}^2 = -\frac{4}{\frac{\pi^4}{\ell^4}\left(\frac{\pi^2}{6}-\frac{15}{16}\right)\sqrt{\frac{EI}{m}}}\left(\frac{\sigma_r}{2} \pm \sqrt{\frac{1}{mEI}\left(\frac{\mu\breve{P}}{4}\right)^2 - \frac{1}{4m^2}\left(c_e + \frac{\pi^4}{\ell^4}\eta I\right)^2}\right). \tag{7.114}$$

It should be noticed that, being $c_3 < 0$, the softening case occurs, as depicted in Fig. 3.4b, d. □

7.6 Dynamic Bifurcation

As an example of Hopf bifurcation triggered by follower forces, the classic Beck beam is considered (Fig. 7.6), for which the critical value of the force and the amplitude of the limit cycle are determined.

Problem 7.10 (*Dynamic bifurcation of the damped Beck beam*) Consider a viscoelastic clamped-free beam, loaded at the tip by a follower force of intensity μF (Fig. 7.6). (a) Write the equation of motion. (b) Apply the MSM to analyze dynamic bifurcations. (c) Determine the critical value μ_c of the load multiplier at which a Hopf bifurcation occurs, as well as the right and left associated eigenvectors. (d) Derive the bifurcation equation and, by using the results of Chap. 4, determine the limit cycle rising at the bifurcation point.

(a) Since the beam is axially unrestrained, the model of Eq. 5.18 apply, namely:

$$\begin{aligned} m\ddot{v} + c_e\dot{v} + EI\left\{v'''' + \left[v'\left(v'v''\right)'\right]'\right\} + \eta I\dot{v}'''' \\ +\mu F v''\left(1+\frac{3}{2}v'^2-\frac{1}{2}v'^2(\ell,t)\right)+\frac{1}{2}m\left[v'\int_\ell^s\left(\int_0^s\left(v'^2\right)^{\cdot\cdot}\mathrm{d}s\right)\mathrm{d}s\right]' = 0, \end{aligned} \tag{7.115}$$

in which $H_\ell \simeq -\mu F\left(1-\frac{1}{2}v'^2(\ell,t)\right)$ and $p_y = 0$ have been replaced. Concerning the boundary conditions, $v(0,t) = v'(0,t) = 0$ hold at the clamped end, and $EIv''(\ell,t)\left(1+\frac{1}{2}v'^2(\ell,t)\right) + \eta I\dot{v}''(\ell,t) = 0$, $-EIv'''(\ell,t)\left(1+\frac{1}{2}v'^2(\ell,t)\right) - \eta I\dot{v}'''(\ell,t) - EIv'(\ell,t)\,v''^2(\ell,t) = 0$ at the free end, according to the discussion in Sect. 5.5.2.

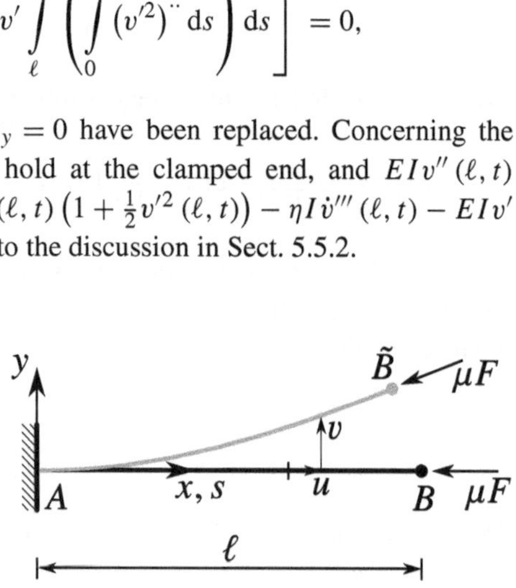

Fig. 7.6 Dynamic bifurcation of the Beck beam

(b) To use the MSM, the rescaling $v \to \epsilon^{1/2}\hat{v}$ is performed, while all damping terms are kept of order 1. Moreover, the bifurcation parameter is split as $\mu = \mu_0 + \epsilon\tilde{\mu}$. Independent time-scales $t_0 := t, t_1 := \epsilon t$ are introduced, for which $\dot{v} = (\partial_0 + \epsilon\partial_1 + \cdots) v, \ddot{v} = (\partial_0^2 + 2\epsilon\partial_0\partial_1 + \cdots) v$. Furthermore, the displacement is expanded as: $v = v_0 + \epsilon v_1 + \cdots$. The following perturbation equations are derived:

Order ϵ^0:

$$m\partial_0^2 v_0 + c_e\partial_0 v_0 + EIv_0'''' + \eta I\partial_0 v_0'''' + \mu_0 F v_0'' = 0. \tag{7.116}$$

Order ϵ^1:

$$\begin{aligned} m\partial_0^2 v_1 + c_e\partial_0 v_1 + EIv_1'''' + \eta I\partial_0 v_1'''' + \mu_0 F v_1'' = & -2m\partial_0\partial_1 v_0 - c_e\partial_1 v_0 - \eta I\partial_1 v_0'''' \\ & - EI\left[v_0'\left(v_0'v_0''\right)'\right]' - \tilde{\mu}Fv_0'' \\ & - \mu_0 F v_0''\left(\frac{3}{2}v_0'^2 - \frac{1}{2}v_0'^2(\ell, t)\right) \\ & - \frac{1}{2}m\left[v_0'\int_\ell^s\left(\int_0^s \partial_0^2\left(v_0'^2\right) \mathrm{d}s\right)\mathrm{d}s\right]'. \end{aligned} \tag{7.117}$$

The relevant geometric boundary conditions are $v_k(0, t_k) = v_k'(0, t_k) = 0, k = 0, 1, \ldots,$ and the mechanical conditions:

Order ϵ^0:

$$EIv_0''(\ell, t_k) + \eta I\partial_0 v_0''(\ell, t_k) = 0, \tag{7.118a}$$

$$-EIv_0'''(\ell, t_k) - \eta I\partial_0 v_0'''(\ell, t_k) = 0. \tag{7.118b}$$

Order ϵ^1:

$$EIv_1''(\ell, t_k) + \eta I\partial_0 v_1''(\ell, t_k) = -\frac{1}{2}EIv_0''(\ell, t_k)\, v_0'^2(\ell, t_k) - \eta I\partial_1 v_0''(\ell, t_k), \tag{7.119a}$$

$$\begin{aligned} -EIv_1'''(\ell, t_k) - \eta I\partial_0 v_1'''(\ell, t_k) = & \frac{1}{2}EIv_0'''(\ell, t_k)\, v_0'^2(\ell, t_k) \\ & + EIv_0''(\ell, t_k)^2\, v_0'(\ell, t_k) + \eta I\partial_1 v_0'''(\ell, t_k). \end{aligned} \tag{7.119b}$$

To normalize the solution, the ω_r-harmonic component of the response $v(\ell, t)$ is imposed to be equal to $a(t)$.

(c) At the (unknown) Hopf bifurcation point $\mu_0 = \mu_c$, the leading order motion is harmonic of (unknown) frequency ω_c on the t_0 scale, i.e.:

$$v_0 = A(t_1, \ldots)\,\phi_c(s)\, e^{i\omega_c t_0} + \text{c.c.}. \tag{7.120}$$

By replacing this solution in the ϵ^0 perturbation Eq. 7.117, a boundary value problem, with complex coefficients, follows, i.e.:

$$
\begin{aligned}
(EI + i\eta I\omega_c)\,\phi_c'''' + \mu_c F \phi_c'' + \left(ic_e\omega_c - m\omega_c^2\right)\phi_c &= 0, && (7.121a)\\
\phi_c(0) = \phi_c'(0) &= 0, && (7.121b)\\
\phi_c''(\ell) = -\phi_c'''(\ell) &= 0, && (7.121c)
\end{aligned}
$$

in which μ_c is the lowest real root. The solution reads:

$$
\phi_c = C_1 \cos(\alpha s) + C_2 \sin(\alpha s) + C_3 \cosh(\beta s) + C_4 \sinh(\beta s), \tag{7.122}
$$

where C_k are arbitrary constants, and:

$$
\begin{aligned}
\alpha^2 &:= \frac{\mu_c F + \sqrt{\mu_c^2 F^2 - 4i\omega_c(c_e + i\omega_c m)(EI + i\eta I\omega_c)}}{2(EI + i\eta I\omega_c)},\\
\beta^2 &:= \frac{-\mu_c F + \sqrt{\mu_c^2 F^2 - 4i\omega_c(c_e + i\omega_c m)(EI + i\eta I\omega_c)}}{2(EI + i\eta I\omega_c)}.
\end{aligned} \tag{7.123}
$$

By enforcing the boundary conditions, a homogeneous problem in the C_k unknowns is derived. By requiring the determinant of the square matrix is zero, a complex characteristic equation is drawn, which splits in two real equations for the two unknowns ω_c and μ_c. As a numerical example, by letting $c_e = 0.1\frac{\sqrt{m\,EI}}{\ell^2}$, $\eta I = 0.01\ell^2\sqrt{m\,EI}$, the following solution is found[10]: $\mu_c = 12.9274\frac{EI}{F\ell^2}$, $\omega_c = 5.923\sqrt{\frac{EI}{m\ell^4}}$; the wave numbers α, β and the constants C_k of the mode, normalized in such a way $\phi_c(\ell) = 1$, are given in Table 7.2.

Since the mode ϕ_c is complex, also the left eigenvector ψ_c is needed to enforce solvability. This is obtained by solving the adjoint problem (see the Appendix A):

$$
\begin{aligned}
(EI - i\eta I\omega_c)\,\psi_c'''' + \mu_c F \psi_c'' + \left(-ic_e\omega_c - m\omega_c^2\right)\psi_c &= 0, && (7.124a)\\
\psi_c(0) = \psi_c'(0) &= 0, && (7.124b)\\
(EI - i\eta I\omega_c)\,\psi_c''(\ell) + \mu_c F \psi_c(\ell) &= 0, && (7.124c)\\
-(EI - i\eta I\omega_c)\,\psi_c'''(\ell) - \mu_c F \psi_c'(\ell) &= 0. && (7.124d)
\end{aligned}
$$

The left eigenvector has the same form than the right one, in which, however, the wave-numbers α, β are redefined as follows:

$$
\begin{aligned}
\alpha^2 &:= \frac{\mu_c F + \sqrt{\mu_c^2 F^2 + 4i\omega_c(c_e - i\omega_c m)(EI - i\eta I\omega_c)}}{2(EI - i\eta I\omega_c)},\\
\beta^2 &:= \frac{-\mu_c F + \sqrt{\mu_c^2 F^2 + 4i\omega_c(c_e - i\omega_c m)(EI - i\eta I\omega_c)}}{2(EI - i\eta I\omega_c)}.
\end{aligned} \tag{7.125}
$$

The α, β and the C_k constants are reported in Table 7.2 for the numerical case considered, adopting the normalization $\psi_c(\ell) = 1$.

---

[10] The complete scenario, when the two damping coefficients are varied, is discussed in [2].

Table 7.2 Right and left eigenvectors of the Beck beam when $c_e = 0.1\frac{\sqrt{m\,EI}}{\ell^2}$, $\eta I = 0.01\ell^2\sqrt{m\,EI}$

	Mode	
	ϕ_c	ψ_c
$\alpha\ell$	$3.8982 - 0.1045\,i$	$3.8982 + 0.1045\,i$
$\beta\ell$	$1.5176 - 0.0170\,i$	$1.5176 + 0.0170\,i$
C_1	$-0.0579 - 0.0148\,i$	$-1.8618 - 0.4298\,i$
C_2	$-0.1309 + 0.0021\,i$	$0.7775 + 0.1705\,i$
C_3	$0.0579 + 0.0148\,i$	$1.8618 + 0.4298\,i$
C_4	$0.3363 - 0.0106\,i$	$-1.9907 - 0.4692\,i$

(d) With Eq. 7.120, the ϵ order perturbation Eq. 7.117 becomes:

$$m\partial_0^2 v_1 + c_e\partial_0 v_1 + EIv_1'''' + \eta I\partial_0 v_1'''' + \mu_c F v_1'' = q_{11}(s)\,e^{i\omega_c t_0} + \text{N.R.T.} + \text{c.c.}, \tag{7.126}$$

where:

$$\begin{aligned} q_{11} :=& -2i\omega_c m\phi_c\partial_1 A - c_e\phi_c\partial_1 A - \eta I\phi_c''''\partial_1 A - \tilde{\mu}F\phi_c'' A \\ &+ \mu_c F A^2\bar{A}\left[\frac{1}{2}\bar{\phi}_c''\left(\phi_c'^2(\ell) - 3\phi_c'^2\right) + \phi_c''\left(\phi_c'(\ell)\bar{\phi}_c'(\ell) - 3\phi_c'\bar{\phi}_c'\right)\right] \\ &+ 2m\omega_c^2 A^2\bar{A}\left(\bar{\phi}_c''\int_\ell^s\int_0^{\check{s}}\phi_r'(\hat{s})^2\,\mathrm{d}\hat{s}\,\mathrm{d}\check{s} + \bar{\phi}_c'\int_0^s\phi_r'(\hat{s})^2\,\mathrm{d}\hat{s}\right) \\ &- EIA^2\bar{A}\Big[3\phi_c''^2\bar{\phi}_c'' + \phi_c'\left(2\phi_c''''\bar{\phi}_c' + 4\phi_c'''\bar{\phi}_c'' + \phi_c'\bar{\phi}_c''''\right) \\ &+ 4\phi_c''\left(\phi_c'''\bar{\phi}_c' + \phi_c'\bar{\phi}_c'''\right)\Big] \end{aligned} \tag{7.127}$$

and:

$$EIv_1''(\ell, t_k) + \eta I\partial_0 v_1''(\ell, t_k) = 0, \tag{7.128a}$$

$$-EIv_1'''(\ell, t_k) - \eta I\partial_0 v_1'''(\ell, t_k) = 0, \tag{7.128b}$$

in which the boundary conditions Eq. 7.121c have been taken into account. Solvability requires the resonant terms to be orthogonal to the left eigenvectors, namely $\int_0^\ell \bar{\psi}_c(s)\,q_{11}(s)\,\mathrm{d}s = 0$, from which, after back rescaling, the bifurcation equation is drawn:

$$\dot{A} = c_1\tilde{\mu}A + c_3 A^2\bar{A}, \tag{7.129}$$

where:

$$
\begin{aligned}
c_1 &:= -\frac{F \int_0^\ell \bar{\psi}_c \phi_c'' \,\mathrm{d}s}{(2i\omega_c m + c_e)\int_0^\ell \bar{\psi}_c \phi_c \,\mathrm{d}s + \eta I \int_0^\ell \bar{\psi}_c \phi_c'''' \,\mathrm{d}s}, \\
c_3 &:= \frac{1}{(2i\omega_c m + c_e)\int_0^\ell \bar{\psi}_c \phi_c \,\mathrm{d}s + \eta I \int_0^\ell \bar{\psi}_c \phi_c'''' \,\mathrm{d}s}\left\{\mu_c F \int_0^\ell \bar{\psi}_c \left[\frac{1}{2}\bar{\phi}_c'' \left(\phi_c'^2(\ell) - 3\phi_c'^2\right)\right.\right. \\
&\quad \left. + \phi_c'' \left(\phi_c'(\ell)\bar{\phi}_c'(\ell) - 3\phi_c'\bar{\phi}_c'\right)\right] \mathrm{d}s + 2m\omega_c^2\left[\int_0^\ell \bar{\psi}_c(s)\,\bar{\phi}_r'(s) \int_0^s \phi_r'(\hat{s})^2 \,\mathrm{d}\hat{s}\,\mathrm{d}s \right. \\
&\quad \left. + \int_0^\ell \bar{\psi}_c(s)\,\bar{\phi}_r''(s) \int_\ell^s \int_0^{\check{s}} \phi_r'(\hat{s})^2 \,\mathrm{d}\hat{s}\,\mathrm{d}\check{s}\,\mathrm{d}s\right] - EI \int_0^\ell \bar{\psi}_c(s)\left[3\phi_c''^2\bar{\phi}_c'' \right. \\
&\quad \left.\left. + \phi_c'\left(2\phi_c''''\bar{\phi}_c' + 4\phi_c'''\bar{\phi}_c'' + \phi_c'\bar{\phi}_c''''\right) + 4\phi_c''\left(\phi_c'''\bar{\phi}_c' + \phi_c'\bar{\phi}_c'''\right)\right] \mathrm{d}s\right\}.
\end{aligned}
\tag{7.130}
$$

In the numerical example considered, it is:

$$
c_1 := (0.0335 + 0.2911\,i)\frac{F}{\sqrt{mEI}}, \quad c_3 := -(0.9203 + 2.5242\,i)\frac{1}{\ell^4}\sqrt{\frac{EI}{m}}. \tag{7.131}
$$

By remembering the results of Chap. 4, and denoting with the subscript R or I the real and imaginary part of the constants c_1, c_3, the amplitude of the limit cycle is:

$$
a_{re} = 2\sqrt{-\frac{c_{1R}}{c_{3R}}(\mu - \mu_c)} = 0.3816\,\ell\sqrt{\frac{F\ell^2}{EI}(\mu - \mu_c)}. \tag{7.132}
$$

Since $c_{1R}/c_{3R} < 0$, the bifurcation is supercritical, as represented in Fig. 4.1a. □

7.7 Summary of the Bifurcation Equations

The bifurcation equations provided by the MSM in all the problems addressed in this chapter are resumed and their form discussed.

Problem 7.11 (*Comparison among the bifurcation equations*) (a) List the complex bifurcation equations, obtained by the MSM, ruling the slow dynamics in the following problems: (i) undamped free vibrations, (ii) undamped and damped forced vibrations in primary external resonance, (iii) forced vibrations in sub-harmonic resonance, (iv) forced vibrations in super-harmonic resonance, (v) principal parameteric excitation, (vi) dynamic bifurcations of autonomous systems. (b) Find and comment a 'rule' to justify the structure of the complex bifurcation equations.

(a) The following bifurcation equations have been found (index r suppressed, all the constant c_k real, except for case (vi), in which they are complex):

(i) Undamped free vibrations (Eq. 7.51):

$$\dot{A} = ic_3 A^2 \bar{A}. \tag{7.133}$$

(ii) Primary external resonance, $\Omega \simeq \omega$. If the system is undamped (Eq. 7.66):

$$\dot{A} = ic_3 A^2 \bar{A} + ic_0 \alpha e^{i\sigma t}; \tag{7.134}$$

if it is damped (Eq. 7.82)

$$\dot{A} = -c_1 A + ic_3 A^2 \bar{A} + ic_0 \alpha e^{i\sigma t}. \tag{7.135}$$

(iii) Sub-harmonic resonance, $\Omega \simeq 3\omega$ (Eq. 7.93):

$$\dot{A} = -c_1 A + ic_3 A^2 \bar{A} + ic_{21} \alpha^2 A + ic_{12} \alpha \bar{A}^2 e^{i\sigma t}. \tag{7.136}$$

(iv) Super-harmonic resonance, $\Omega \simeq \frac{1}{3}\omega$ (Eq. 7.96):

$$\dot{A} = -c_1 A + ic_3 A^2 \bar{A} + ic_{21} \alpha^2 A + ic_{30} \alpha^3 e^{i\sigma t}. \tag{7.137}$$

(v) Principal parametric excitation (Eq. 7.111):

$$\dot{A} = -c_1 A + ic_3 A^2 \bar{A} + ic_0 \mu \bar{A}_r e^{i\sigma t}. \tag{7.138}$$

(vi) Dynamic (Hopf) bifurcation (Eq. 7.129):

$$\dot{A} = c_1 \tilde{\mu} A + c_3 A^2 \bar{A}. \tag{7.139}$$

(b) The following 'rule' can be used to check (or even to foresee) the structure of monomials appearing on the right side of the bifurcation equation: (i) multiply A by the exponential $e^{i\omega t}$, $\bar{A}$ by the exponential $e^{-i\omega t}$, μ by $e^{i\Omega_p t}$ and α by $e^{i\Omega_f t}$; (ii) replace Ω_f (the frequency of the forcing) by $\pm\omega$, $\pm 3\omega$, $\pm\frac{1}{3}\omega$, according to the nature of the excitation, namely primary, sub- or super-harmonic, respectively[11]; (iii) replace Ω_p (the frequency of parametric excitation) by 2ω; (iv) check that the resulting exponential of all the monomials is equal to $e^{i\omega t}$ (which appears to the left of the equation). The rule is a obvious consequence of having selected the resonant terms in the perturburbation process. The amplitude of motion A, the amplitude of the forcing α and the amplitude of the pulsating force μ are 'labels' which remember the frequency associated with each of them.[12] □

[11] The $\pm$ sign is necessary to tackle the α^2 terms in Eqs. 7.136, 7.137, whose associated frequency, indeed, should be taken zero.

[12] It is not convenient to refer to σ instead of α, since the former depends on how it has been introduced. For example, the equivalent forms $3\Omega = \omega + \sigma$ or $\Omega = \frac{1}{3}\omega + \sigma$ attribute to σ different meanings.

References

1. Irvine, M.: Cable Structures. Dover, New York (1992)
2. Luongo, A., Ferretti, M., Di Nino, S.: Stability and Bifurcation of Structures: Statical and Dynamical Systems. Springer, Cham (2023)
3. Luongo, A., Rega, G., Vestroni, F.: On nonlinear dynamics of planar shear indeformable beams. J. Appl. Mech. **53**(31–32), 619–624 (1986)
4. Pignataro, M., Rizzi, N., Luongo, A.: Stability, Bifurcation and Postcritical Behaviour of Elastic Structures. Elsevier, Amsterdam (1990)

8 Summary

8.1 Introduction

It has been shown, in this book, that any visco-elastic continuous structure can be tackled by proper perturbation methods, able to reduce the infinite-dimensional space to a finite space of small dimensions (e.g., 1 or 2). The reduction process is carried out in systematic way by (a) the *straightforward expansion* or (b) the *strained parameter perturbation method*, in statics, and by (c) the *Multiple Scale Method*, in dynamics. All problems are reduced, by series expansions, to a sequence of linear problems governed by the same differential operator, to be solved in chain. If the operator is singular, compatibility conditions are required at each steps, leading to build-up the desired bifurcation equations. The dimension of these equations is equal to the number of free quantities contained in the generating solution (typically, the amplitude and phase), and therefore it is equal to the dimension of the critical (or center) subspace of the linear problem.

In this chapter, the nonlinear models are rivisited and the algorithms and results summarized for each class of problems examined.

8.2 Nonlinear Models

Nonlinear phenomena call for formulating and using nonlinear models. The focus here is on *continuous*, rather than discrete, systems. The simplest of them is a one-dimensional model, in which nonlinearities are all of algebraic type, namely a *linear* Euler-Bernoulli beam resting on *nonlinear* Winkler soil, of polynomial type. This system, when suitably loaded, is able to display any of the phenomena of interest in this book. In the static regime: (i) when the beam is transversely loaded, it exhibits a nonlinear elastostatic response, (ii) when it is axially compressed by a gravitational force, it manifests *buckling* and postbuckling. In the

A. Luongo et al., *Perturbation Methods and Nonlinear Phenomena*,
Synthesis Lectures on Engineering, Science, and Technology,
https://doi.org/10.1007/978-3-031-49397-3_8

dynamic regime: (i) when transverse loads, harmonically varying in time, are applied to the beam, the *external resonance* phenomenon takes place; (ii) when the beam is compressed by a harmonically varying axial force, the *parametric resonance* phenomenon occurs; (iii) when a compression *follower force* is applied at the free end of the beam, a *dynamic* (or Hopf) *bifurcation* manifests; (iv) if a gravitational compressive force is added to the follower force, *static bifurcations* also occur for nonconservative systems.

More complex (and realistic) nonlinear elastic systems have been addressed. They all are encompassed by a *metamodel*, which is expressed in terms of *abstract* linear and nonlinear differential (or integro-differential) operators, defined in the domain and at the boundary, that act on the configuration variables to transform them into forces, which enter the balance equations. The metamodel is useful to illustrate algorithms, since avoids tedious repetitions relevant to the specific systems, highlighting, in contrast, the common aspects.

Several one-dimensional and two-dimensional nonlinear models of structural elements have been discussed [3–5, 9]. These have been named *minimal models*, since they contain only the essential nonlinearities, able to qualitatively capture the phenomenon, although they could be affected by quantitative errors, due to the fact they are somewhat inconsistent (in the sense that terms of the same order of those retained, are instead disregarded). Concerning strings, cables and beams longitudinally constrained at the ends, the longitudinal displacements have been *statically condensed*, by neglecting the relevant inertia forces. The procedure leads to integro-differential equations in the transverse displacements only. In contrast, when a membrane or a plate with constrained in-plane displacements at the boundary are considered, even neglecting in-plane inertia, the static condensation is no longer viable, so that the models consist of a dynamic transverse equation coupled to two static in-plane equations.

Continuous models can, of, course, be discretize by using the *Galerkin approach*, which consists in expressing the response in a basis of suitably chosen trial functions. However, due to the nonlinear nature of the problem, a sufficient number (not a priori known) of them must be taken, in order to capture the modification of the static or dynamic shape of the deflection, when the amplitude of the response increases (as it is well explained by the perturbation approach). Just in the case of symmetric systems, this modification of the deflection is a higher-order effect, and therefore using the linear response as a unique trial function gives acceptable results.

8.3 Algorithms and Results

In all cases dealt with, a perturbation method has been used, which consists of an extrapolation of the solution from the associated linear problem. The method calls for the following common steps to be performed [8]:

1. introduce in the equations a perturbation parameter ϵ (to be removed at the end of the procedure), by rescaling the state-variable as $\mathbf{u} \rightarrow \epsilon\hat{\mathbf{u}}$, if the system is non-symmetric, and $\mathbf{u} \rightarrow \epsilon^{1/2}\hat{\mathbf{u}}$, if it is symmetric;
2. expand the variable in series of powers of ϵ, substitute it in the field equation and boundary conditions, separately equate to zero the terms with the same power of ϵ, and obtain the perturbation equations;
3. solve in sequence these perturbation equations; they are all linear and share the same field- and boundary-operator, thus obtaining the coefficients of the series.

However, the five classes of problems mentioned above (except the first one), need further specific steps, resumed ahead.

Elastostatics

In static analysis, a *straightforward expansion* is adopted [8]. The generating equation is non-homogeneous, and the differential operator (including boundary conditions) is non-singular, so that each perturbation equation admits a unique solution. The load-displacement curve is tangent at the origin to the straight line of the linear theory.

Buckling

In buckling analysis, the *strained parameter method* is used [8]. It consists in expanding, in addition to the variable $\mathbf{u}$, also the prestress multiplier ν around an unknown critical value ν_c. Since the generating problem is homogeneous, ν_c is used as an eigenvalue to solve a space eigenproblem, rendering the operator singular. Since the solutions are infinite in number, the associated eigenvector is normalized, and an unknown amplitude a is introduced as a factor. By going on to higher orders, since the operator is singular, compatibility conditions must be systematically enforced, requiring orthogonality of the right hand side (including forces at the boundary) to the critical eigenvector. These conditions provide the equations useful to determine the values of the coefficients of the series expansion of ν. The series describes the mechanical behavior of the beam on the (ν, a)-plane. If the system is non-symmetric, a *transcritical* bifurcation occurs; if it is symmetric, a *pitchfork* bifurcation takes place. However, these are ideal bifurcations, relevant to perfect systems; if, in contrast, small imperfections exist, they must be properly rescaled in the perturbation scheme, and the branch points disappear, while nontrivial equilibrium paths exist, possibly exhibiting *limit points*, occurring at loads smaller than the critical load. The strained parameter method is unable to give an answer concerning the stability of the equilibrium.

Nonlinear External Resonances

In nonlinear dynamic analysis, the *Multiple Scale Method* (MSM) is applied [7, 8]. It requires introducing several independent time scales $t_k = \epsilon^k t$, according to the idea that periodic oscillations of linear systems are slowly-modulated in amplitude and phase, when nonlinearities are introduced. Some rescaling are required for all the small quantities, namely: the damping coefficient, the amplitude of the external excitation (when in primary resonance), the detuning between its frequency and a natural frequency of the beam. Rescaling is made in such a way to order all these terms at the same level at which the most significant nonlinearities appear first, namely the cubic ones, since the quadratic terms do not create resonant frequencies. Equations are solved in chain, starting from the homogeneous generating equation.

In the primary resonance case, in which the excitation frequency Ω is close to a natural one, ω_r, the (nondimensional) amplitude of the excitation must be considered small (*soft excitation*), since it triggers a (nondimensional) response much larger than the force itself, as highlighted by the linear problem, in which the response tends to infinite. The lower order (also said generating) solution is a harmonic motion on the fast scale, in which only the resonant mode is considered, if *internal resonance* conditions are excluded. Compatibility at each step (stating that the resonant terms must be orthogonal to the rth natural mode) furnishes first-order differential equations, ruling the evolution of amplitude and phase of the mode on progressively slower scales. Differently from the strained parameter method, in which these equations were solved at each steps, in the MSM they are set aside, in order to be recombined in a whole at the end of the procedure. As a result, a finite-dimensional dynamical system equivalent to the infinite-dimensional original system is found, capturing the long-term essential dynamics of the beam. Therefore the MSM acts as a *reduction method*[1]; moreover, it filters the fast dynamics, so that, numerical integrations are much more easier to be performed. The bifurcation equation permits evaluation of the frequency-response law, i.e., the relationship between the amplitude of periodic solutions (limit cycles) admitted by the system and the excitation frequency. Moreover, decides about the stability of the periodic motions.

Finite-amplitude *free oscillations* have also been studied, as limit for the external excitation intensity tending to zero; thus, the concepts of *backbone curve* of the frequency response law is introduced.

The primary resonance, however, does not represent the unique possible nonlinear external resonance. Indeed, when the excitation frequency is a multiple or sub-multiple of a natural frequency (e.g., $\Omega \simeq 3\omega_r$, or $3\Omega \simeq \omega_r$) and the excitation amplitude is large enough, a sub-harmonic or super-harmonic resonance, correspondingly, occurs. It is generated by the following mechanism, well highlighted by the perturbation method. If the excitation amplitude is large (*hard excitation*), it should not be rescaled, but must appear as a know term in the generating equation. Therefore, the lower order solution contains two frequencies: ω_r in

[1] The MSM supplies the same results of the Center Manifold and Normal Form Theory [6, 11].

the complementary component, and Ω in the particular part. These basic frequencies, when substituted in the higher order perturbation equations, generate new frequencies which are combinations of the two basic ones, which could reproduce the natural frequency, thus causing resonances. For example, cubic nonlinearities produce the forcing frequencies $\Omega - 2\omega_r$ and 3Ω, which, in the two cases mentioned above, are indeed resonant. As for the primary resonance, a specific bifurcation equation is derived for each of the two resonances.

Parametric Excitation

A different type of excitation, caused by harmonic loads, is the parametric one. It occurs when the external load (or a prescribed motion of the constraints) induces a periodic variation of the mechanical properties of the system. As an example, when a (Bolotin) beam is loaded by an axially pulsating load, of frequency Ω, the geometric stiffness of the beam is time-varying. Such a problem has been tackled by the MSM by dealing with symmetric systems only, by shifting the parametric excitation at the same level in which the (cubic) nonlinearities and damping first appear. The generating solution, accordingly, and in absence of internal resonances, only contains the natural frequency ω_r of the mode involved in the parametric resonance. Such a frequency combines itself with the parametric excitation to give rise to the frequency $\Omega - \omega_r$, while producing, as usual for cubic nonlinearities, the resonant frequency ω_r. The former of them, related to the linear terms, is also resonant if $\Omega = 2\omega_r$ (said *principal parametric resonance*), so that the parametric excitation competes with nonlinear terms. It is concluded that, if the parametric frequency is double of a natural frequency, the trivial equilibrium configuration loses stability. Accordingly, the linear response would tend to infinity, but, such a response is limited by the nonlinearities, which cause the birth of *limit cycles*.

The bifurcation equation provided by the MSM permits to analyze the neighborhood of the perfect parametric resonance. It is found that there exist a region, in the excitation amplitude-frequency parameter plane, in which instability manifests itself. The boundaries of the unstable regions are the geometric locus at which the limit cycles bifurcate. From the same bifurcation equation, the amplitude of the limit cycle is determined as a function of the two bifurcation parameters, and information about stability is drawn.

Dynamic Bifurcation

Dynamic (or Hopf) bifurcation is suffered by nonconservative systems, as the (Beck) beam loaded by a follower force. The relevant analysis is carried out by the MSM. Differently from the former problems, here damping is *not* rescaled, so that it appears in the generating equation; thus, coalescence of eigenvalues, more difficult to be tackled, is avoided. In contrast, the prestress intensity μ is split in an unknown critical part μ_c, and in an incremental part $\tilde{\mu}$ (similarly to what done for the frequency of the external excitation). The critical part is still found as solution of a space eigenproblem, derived from the generating partial differential equation, once harmonic dependence on time of the solution has been enforced. The incremental part, instead, appears in the differential equations which express compatibility at various orders, and, from these, in the reconstituted *bifurcation equation* which collect them. This equation governs the long-term dynamics of amplitude and phase. Therefore, the original continuous system is reduced to a two-dimensional dynamical system, according to the *Center Manifold Theory*. It states that, close to a bifurcation, the dynamics develop on some manifold of the state-space, whose dimension equates that of the critical subspace of the linearized system (namely two, in Hopf bifurcation), to which it is tangent at the origin [1, 2, 10, 12]. The study of the bifurcation equations leads to detect the existence of a family of *limit cycles* that, if supercritical, are stable and avoid the blow-up of the linear solution. The Hopf bifurcation is insensitive to imperfections.

Static Bifurcation of Dynamical Systems

Static (or divergence) bifurcations, can also been studied in the framework of dynamics, thus getting information not only on equilibrium, but also on stability. It has been shown how to apply the MSM to a nonconservative beam suffering static bifurcation. The procedure closely repeats that followed for dynamic bifurcation, with the only difference that the generating equation does not admit now a harmonic motion, but a constant solution on the fast scale, slowly modulated on the slow scales. Therefore, 'resonant terms' are constant on the fast scale, and a one-dimensional bifurcation equation is got, in the only unknown amplitude (since there is no phase). This equation furnishes the same equilibria detectable by the strained parameter method, but it determines also the stability character of the branched equilibria.

Role of Symmetry

Comparing non-symmetric and symmetric systems, it has been stressed that quadratic nonlinearities appearing in the former systems are responsible for change of the spatial shape in the evolutionary process (relevant to both static and dynamic problems). Therefore, the deformed pattern of the system is not only affected by a scalar factor, as it happens in the linear theory, but it modifies itself during the evolution. This is due to the fact that *passive modes*, which are not directly involved in instability/resonance (of static or dynamic type), do participate to the response, as driven by the active mode. Since the perturbation method is able to express this dependence, the final equations are low-dimensional. In contrast, if quadratic nonlinearities are absent, the first meaningful information (first-order asymptotic solution) is obtained *before* such a shape modification appears in the perturbation scheme.

This circumstance entails that, for symmetric systems, the change in the response of the spatial shape is not so important as for non-symmetric systems.

Discrete Systems

Finally, the Galerkin approach has been illustrated, leading to ordinary time-differential equations, instead of partial differential equations. The perturbation algorithms apply to discrete systems practically in the same form as to continuous systems. The trial functions must be chosen in order to describe linear as well as nonlinear responses. However, for symmetric systems, a single degree of freedom system gives the same solution of the continuous approach, to within the truncation adopted.

References

1. Carr, J.: Applications of Centre Manifold Theory. Springer, New York (1981)
2. Guckenheimer, J., Holmes, P.: Nonlinear Oscillations, Dynamical Systems, and Bifurcations of Vector Fields. Springer, New York (1983)
3. Irvine, M.: Cable Structures. Dover, New York (1992)
4. Lacarbonara, W.: Nonlinear Structural Mechanics: Theory, Dynamical Phenomena and Modeling. Springer Science & Business Media, New York (2013)
5. Luongo, A., Zulli, D.: Mathematical Models of Beams and Cables. Wiley, New York (2013)
6. Nayfeh, A.H.: The Method of Normal Forms. Wiley-VCH, Weinheim (2011)
7. Nayfeh, A.H., Mook, D.T.: Nonlinear Oscillations. Wiley, New York (1995)
8. Nayfeh, A.H.: Perturbation Methods. Wiley, New York (1973)
9. Timoshenko, S.P., Woinowsky-Krieger, S.: Theory of Plates and Shells. McGraw-Hill, New York (1959)
10. Troger, H., Steindl, A.: Nonlinear Stability and Bifurcation Theory: An Introduction for Engineers and Applied Scientists. Springer, Wien (1991)
11. Steindl, A., Troger, H.: Methods for dimension reduction and their application in nonlinear dynamics. Int. J. Solids Struct. **38**(10–13), 2131–2147 (2001)
12. Wiggins, S.: Introduction to Applied Nonlinear Dynamical Systems and Chaos. Springer, New York (1990)

Compatibility Conditions and Normalization

A

A.1 Introductory Remarks

The perturbation methods illustrated in this book transform weakly *nonlinear* problems in a cascade of *linear* problems. In solving them, one encounters situations in which, either: (a) a unique solution exists for *any* known term (this happening when the operator is *non-singular*), or, (b) no solutions exist for an arbitrary know terms, but infinite solutions exist for *special* known terms (occurring when the operator is *singular*). This specialty, involving the *data* of the problem, is expressed by an equation which is known as *compatibility condition*, we want to derive here. The question is well-known in Functional Analysis, but it could be new to the reader, so it is worth discussing it, also by exploiting mechanical arguments. Since we are interested in systems in which the number of equations and unknowns is equal (squared operators), we will confine ourselves to this simpler case.

Moreover, when one has to deal with infinite solutions, it is advisable to have a criterion of *normalization* to chose one among them, as it happens, for example, in linear dynamics, when one normalizes the natural modes. The discussion starts with generic discrete systems. Then, attention is focused on continuous systems, by referring to the linear counterpart of the sample beam in Eqs. 1.7.

A.2 Discrete Systems

To introduce the problem in a plane way, we first recall some results from Linear Algebra, which apply to *discrete* (instead of continuous) systems, as, for example, assemblies of elastically constrained rigid bodies, or elastic frames loaded just at the joints. At a first stage, we consider static problems, by neglecting inertia and damping forces; then, dynamical systems will be considered.

A. Luongo et al., *Perturbation Methods and Nonlinear Phenomena*, Synthesis Lectures on Engineering, Science, and Technology, https://doi.org/10.1007/978-3-031-49397-3

A.2.1 Static Case

For linear static systems, the elastic problem leads to:

$$\mathbf{K}\mathbf{x} = \mathbf{p}, \tag{A.1}$$

where $\mathbf{K}$ is the *square* stiffness matrix, $\mathbf{x}$ the displacement vector and $\mathbf{p}$ the load vector. We put ourselves in a more general context, by allowing $\mathbf{K}$ to be non-symmetric, as it happens when beams are prestressed by nonconservative forces, and equilibrium is enforced in the adjacent configuration. Now, it is well-known from the Rouché-Capelli theorem that Eq. A.1 admits, either: (a) one solution for any $\mathbf{p}$, if $\mathbf{K}$ is non-singular (i.e., if $\det \mathbf{K} \neq 0$, as it occurs for statically-determinate or overdetermined systems); (b) no solutions, if $\mathbf{K}$ is singular (i.e., if $\det \mathbf{K} = 0$, as it occurs in kinematically indeterminate or degenerated systems) and, moreover, $\mathbf{p}$ is generic; (c) infinite solutions, if $\det \mathbf{K} = 0$ but $\mathbf{p}$ satisfies a *compatibility condition*. This condition can be expressed in various forms. May be, the most known of them requires that $\text{rank}\mathbf{K} = \text{rank}\left[\mathbf{K} \mid \mathbf{p}\right]$, but this does not help us in preparing the ground for continuous systems, where the operators are differential and not algebraic. Therefore we use an alternative approach.

Bilinear Identity

Let us consider the so-called *adjoint problem*, defined as:

$$\mathbf{K}^T\mathbf{y} = \mathbf{p}^\star, \tag{A.2}$$

in which $\mathbf{K}^T$ is the transposed stiffness matrix, and $\mathbf{y}$ and $\mathbf{p}^\star$ are dummy variables, associated with $\mathbf{x}$ and $\mathbf{p}$, respectively. If $\mathbf{K} = \mathbf{K}^T$, the original problem in Eq. A.1 is said *self-adjoint*. Equation A.2 is not requested to have a physical meaning, but it should be regarded as an *auxiliary mathematical problem*. An obvious identity holds between the original $\mathbf{K}$ and adjoint $\mathbf{K}^T$ operators, namely:

$$\mathbf{y}^T\mathbf{K}\mathbf{x} = \mathbf{x}^T\mathbf{K}^T\mathbf{y}, \quad \forall \mathbf{x}, \mathbf{y}, \tag{A.3}$$

which is called the *bilinear identity* [1]. From this, it follows straightforwardly that

$$\mathbf{y}^T\mathbf{p} = \mathbf{x}^T\mathbf{p}^\star. \tag{A.4}$$

This latter has, in mechanics, a powerful meaning, since it expresses the equality of the virtual works expended by the two systems of forces, $\mathbf{p}$ and $\mathbf{p}^\star$, in the cross-displacements $\mathbf{y}, \mathbf{x}$, respectively. Therefore it generalizes the *Betti reciprocity theorem* of linear elasticity, which holds for self-adjoint systems only.

Now, if we consider $\mathbf{p}^\star = \mathbf{0}$, it follows that $\mathbf{y}^T\mathbf{p} = \mathbf{0}$. We draw the information that the *range* of $\mathbf{K}$ (i.e., the space spanned by $\mathbf{p}$ when $\mathbf{x}$ spans the domain of $\mathbf{K}$) is orthogonal to the *kernel* of $\mathbf{K}^T$ (i.e., to the null-space spanned by $\mathbf{y}$).[1] Since Eq. A.1 admits a (not necessarily unique) solution if and only if the know term $\mathbf{p}$ is in the range of $\mathbf{K}$ (in such a way there exist some $\mathbf{x}$ whose image is the given $\mathbf{p}$), we conclude that:

> **Compatibility condition:** The Eq. A.1 admits solution if and only if:
>
> $$\mathbf{y}^T\mathbf{p} = \mathbf{0}, \quad \forall \mathbf{y} \mid \mathbf{K}^T\mathbf{y} = \mathbf{0}. \tag{A.5}$$
>
> In words: *in order for a linear problem to be solvable, the known term must be orthogonal to any solutions of the adjoint homogeneous problem.*

Remark A.1 Compatibility requires that the forces $\mathbf{p}$ expend zero virtual work in the displacement (so called floppy-modes) admitted by the adjoint homogeneous system. Of course, if $\mathbf{K}$ (and therefore $\mathbf{K}^T$) is non-singular, then $\mathbf{y} = \mathbf{0}$ (i.e., the kernel of the adjoint is empty), so that the compatibility condition is trivially satisfied by any $\mathbf{p}$.

Normalization Condition

When $\mathbf{K}$ is singular and compatibility has been satisfied, Eq. A.1 admits infinite solutions. By limiting ourselves to the simplest case in which the matrix has lowering in rank equal to 1 (i.e., its kernel is one-dimensional), the solutions read:

$$\mathbf{x} = \mathbf{x}_p + c_h\mathbf{x}_h, \quad \forall c_h, \tag{A.6}$$

where $\mathbf{x}_p$ is a particular solution, $\mathbf{x}_h$ is the complementary solution (i.e., that of the homogeneous problem), here taken of unitary length, $\|\mathbf{x}_h\| = 1$, and c_h is an arbitrary constant. To choose c_h, it is opportune to introduce a *normalization condition*, for example of the type:

$$\mathbf{w}^T\mathbf{x} = a, \tag{A.7}$$

where $\mathbf{w}$ is a properly selected weight-vector, and $a \in \mathbb{R}$ is a constant, assuming the meaning of *amplitude*. Consequently, $c_h = \frac{a-\mathbf{w}^T\mathbf{x}_p}{\mathbf{w}^T\mathbf{x}_h}$. Several choices can be done for $\mathbf{w}$, for example: (a) by taking $\mathbf{w}$ as the kth canonical vector $\mathbf{e}_k$, normalization identifies the amplitude a as the k-th component of $\mathbf{x}$; (b) by taking $\mathbf{w} := \mathbf{x}_h$, the amplitude is meant as the projection of $\mathbf{x}$ on $\mathbf{x}_h$; (c) by taking $\mathbf{w} := \mathbf{y}$, the amplitude is the projection of $\mathbf{x}$ on the kernel of the adjoint operator (i.e., the component of $\mathbf{x}$ orthogonal to the range of $\mathbf{K}$).

---

[1] This property is stated in Algebra by saying that the kernel of the adjoint operator is the *orthogonal supplement* to the range of the operator.

A.2.2 Dynamic Case

Let us extend the previous concepts to linear dynamical systems, governed by:

$$\mathbf{M}\ddot{\mathbf{x}} + \mathbf{C}\dot{\mathbf{x}} + \mathbf{K}\mathbf{x} = \mathbf{p}(t), \tag{A.8}$$

where $\mathbf{M}$ is the inertia (or mass) matrix, $\mathbf{C}$ the damping matrix and $\mathbf{p}(t)$ the load, now depending on time. Although compatibility could be addressed in a more general way, we will confine ourselves to a special case, which will be the unique one we will encounter in this book, namely the *harmonic load*, $\mathbf{p}(t) = \mathbf{q}\exp(i\Omega t)$, where Ω is the frequency, and complex variables are used. We will look for compatibility conditions involving $\mathbf{q}$ and Ω to assure that $\mathbf{x}(t)$ is also harmonic, of the same frequency, namely $\mathbf{x}(t) = \mathbf{u}\exp(i\Omega t)$. By enforcing harmonicity, Eq. A.8 leads to:

$$\mathbf{S}(\Omega)\mathbf{u} = \mathbf{q}, \tag{A.9}$$

where:

$$\mathbf{S}(\Omega) := \mathbf{K} + i\Omega\mathbf{C} - \Omega^2\mathbf{M} \tag{A.10}$$

is the complex, frequency-dependent, *dynamic stiffness matrix*. It should be noticed, that Eq. A.9 is formally equal to the static Eq. A.1, with $\mathbf{K}$ (the static stiffness) replaced by $\mathbf{S}(\Omega)$ (the dynamic, frequency-dependent, stiffness). Thus, even if $\mathbf{K}$ is non-singular, $\mathbf{S}(\Omega)$ can be singular for some Ω (i.e., if Ω equates an undamped natural frequency ω_r, in absence of damping), or quasi-singular (e.g., if $\Omega = \omega_r$ and damping is small). In this case *a resonance* (or quasi-resonance) is said to occur.

Since Eqs. A.9 are complex, it is suitable to define an adjoint operator in the complex field, by using the *Hermitian transposed* matrix $\mathbf{S}^H := (\mathbf{K}^T - i\Omega\mathbf{C}^T - \Omega^2\mathbf{M}^T)$ (transposed conjugate[2]), instead of the simply transposed $\mathbf{S}^T$. Hence, the adjoint problem reads:

$$\mathbf{S}^H\mathbf{v} = \mathbf{q}^\star, \tag{A.11}$$

from which the compatibility condition reads:

$$\mathbf{v}^H\mathbf{q} = \mathbf{0}, \quad \forall\mathbf{v} \,|\, \mathbf{S}^H\mathbf{v} = \mathbf{0}. \tag{A.12}$$

Compatibility expresses the requirement that the loads expend zero virtual work in the (adjoint) resonant mode. If this is not the case, the response is no more harmonic, but diverging in time, of the type $\mathbf{x}(t) = t\,\mathbf{u}\exp(i\Omega t)$, i.e., a circumstance we want to exclude. These mixed algebraic-harmonic terms are also known in literature as *secular terms*, and therefore the locution 'removing secular terms' is very often used as synonymous of 'enforcing compatibility condition'. We will prefer this last wording, to deal consistently with static and dynamical systems.

---

[2] Here, $\mathbf{M} = \mathbf{M}^T$, due to the existence of the kinetic energy.

Concerning normalization, the considerations done for static systems still apply. Solution to Eq. A.9 reads $\mathbf{u} = \mathbf{u}_p + c_h \mathbf{u}_h$, $\forall c_h$, and normalization $\mathbf{w}^H \mathbf{u} = A \in \mathbb{C}$.

A.3 Continuous Static Systems

We address the problem of formulating compatibility conditions for continuous static systems, by generalizing the concepts of the previous section. Instead of considering an abstract system, we refer to the beam resting on Winkler soil, for which the linear static problem follows from Eqs. 1.7, in which inertia, damping, parametric excitation and nonlinear elastic forces are disregarded, namely:

$$u'''' + 2\,(\mu + \nu)\, u'' + k_1 u = \alpha\, p(s), \tag{A.13a}$$

$$u(0) = 0, \quad u'(0) = 0, \tag{A.13b}$$

$$u''(1) = 0, \quad -u'''(1) - 2\nu u'(1) = \alpha P_B, \tag{A.13c}$$

where $u = u(s)$, $P_B = \text{const}$ and $\nu_s = \nu$.

Green (Bilinear) Identity

To write a bilinear identity, of the type of Eq. A.3, we follow these steps [1]: (1) we multiply the field equation, *rendered homogeneous*, by a (sufficiently smooth) dummy function $v = v(s)$; (2) we integrate it on the domain and use integration by parts as many times as needed to 'free' the true variable $u = u(s)$ from the derivatives, by moving them to $v(s)$; namely:

$$\begin{aligned} \int_0^1 v(u'''' + 2(\mu + \nu)u'' + k_1 u)\, \mathrm{d}s &= \int_0^1 u(v'''' + 2(\mu + \nu)v'' + k_1 v)\, \mathrm{d}s \\ &+ [v u''' - v' u'' + (v'' + 2(\mu + \nu)v)u' - (v''' + 2(\mu + \nu)v')u]_0^1. \end{aligned} \tag{A.14}$$

This is called the *Extended Green Identity*. It differs from its algebraic counterpart, Eq. A.3, not only for the obvious fact that now we are dealing with differential operators, so that the scalar product entails the presence of integrals, but mainly for the appearance of boundary terms, generated by integration by parts. However, we can request that boundary terms vanish separately, in order to transform Eq. A.14 in an equality between two integrals, called the *Green Identity*. This identity furnishes the *adjoint field equation* (see the Eq. A.15a below), while vanishing of boundary terms supplies the *adjoint boundary conditions* (see the Eqs. A.15b, c below). These latter are obtained by accounting for the original boundary conditions Eqs. A.13b, c, also *rendered homogeneous*, and arbitrariness of $u''(0)$, $u'''(0)$, $u(1)$, $u'(1)$, which are *not* prescribed by the original boundary conditions. By summarizing, the *adjoint homogeneous problem* is found:

$$v'''' + 2(\mu + \nu)v'' + k_1 v = 0, \tag{A.15a}$$

$$v(0) = 0, \quad v'(0) = 0, \tag{A.15b}$$

$$v''(1) + 2\mu v(1) = 0, \quad -v'''(1) - 2(\mu + \nu)v'(1) = 0. \tag{A.15c}$$

It follows from Eqs. A.15, when compared with Eqs. A.13, that the field equation (or, equivalently, the linear differential operator) is *self-adjoint*, since it is identical to its adjoint; however, boundary conditions are *not self-adjoint*, since they differ in the terms proportional to the follower force μ. Therefore the problem, in a whole, is not-self adjoint, as a consequence of the presence of the nonconservative force. If this is absent, the problem becomes self-adjoint. Thus, conservativeness and self-adjointness are synonymous.

Compatibility Condition

Once the adjoint problem has been built-up, the compatibility condition is easily found. By considering displacements $u(s)$ now satisfying the *non-homogeneous* Eqs. A.13, and displacements $v(s)$ belonging to the set $\mathcal{V}$ of all function satisfying Eqs. A.15, it follows from the extended Green identity, that:

$$\int_0^1 v(s)p(s)\,\mathrm{d}s + v(1)P_B = 0, \quad \forall\, v(s) \in \mathcal{V}. \tag{A.16}$$

Thus, a result formally analogous to Eq. A.5 is recovered, i.e.:

> **Compatibility condition:** *The Eqs.* A.13 *admit solution if and only if the known term is orthogonal to all the solution of the adjoint homogeneous problem, Eqs.* A.15.

Remark A.2 Compatibility Eq. A.16 has a clear mechanical meaning: it requires that the external forces, in the domain $p(s)$ *and* on the boundary P_B, expend, as a whole, zero virtual work in any displacement field admitted by the adjoint homogeneous system (floppy modes).

Normalization Condition

If the differential operator is singular (and, for simplicity, we admit that its kernel is one-dimensional), after compatibility has been satisfied, Eqs. A.13 admit infinite solutions:

$$u(s) = u_p(s) + c_h u_h(s), \quad \forall c_h, \tag{A.17}$$

where u_p is a particular solution and $u_h(s)$ is the solution to the homogeneous problem, e.g., normalized as $\int_0^1 u_h^2(s)\,\mathrm{d}s = 1$. Normalization now requires introducing a linear functional $\mathcal{L}(u)$, which operates on $u(s)$ to give back a scalar. Thus, we write normalization as:

$$\mathcal{L}(u) := \int_0^1 w(s)u(s)\,\mathrm{d}s = a, \tag{A.18}$$

where $w(s)$ is a weighting function. For example, by taking for w the Dirac delta centered at the abscissa s^*, i.e., $w(s) := \delta(s - s^*)$, the amplitude is identified as the displacement at that abscissa, namely $a = u(s^*)$. Other choices, as $w(s) := u_h(s)$, or $w(s) := v(s)$ are also usually made.

Remark A.3 Note that orthogonality involves active forces both in the domain and on the boundary. In a more general case, in which also the displacements are prescribed at the constraints, the compatibility condition involves these known terms, too. In particular, it can be checked that compatibility expresses the equality between: (a) the virtual work spent by all (active) true forces, in the domain and on the boundaries, over the displacements of the adjoint homogeneous problem, and, (b) the virtual work spent by the constraint reactions of the adjoint homogeneous problem over the prescribed displacements of the true system. An exemplification of this property is given in the Exercise A.1.

A.4 Continuous Dynamical Systems

The dynamic problem is now addressed for the beam on elastic soil. Nonlinear Eqs. 1.7 with $\nu_s = \nu$ and $\nu_d = 0$, when linearized, reduce to:

$$\ddot{u} + \xi\dot{u} + u'''' + 2(\mu + \nu)u'' + k_1 u = \alpha p(s,t), \tag{A.19a}$$

$$u(0,t) = 0, \quad u'(0,t) = 0, \tag{A.19b}$$

$$u''(1,t) = 0, \quad -u'''(1,t) - 2\nu u'(1,t) = \alpha P_B, \tag{A.19c}$$

where $u = u(s,t)$. By considering time harmonic loads $p(s,t) = q(s)\exp(i\Omega t)$, $P_B(t) = Q_B\exp(i\Omega t)$, with $q(s)$ and Q_B complex quantities[3] and looking for responses of type $u(s,t) = \phi(s)\exp(i\Omega t)$, the previous equations furnish:

$$\phi'''' + 2(\mu + \nu)\phi'' + (k_1 + i\Omega\xi - \Omega^2)\phi = \alpha q(s), \tag{A.20a}$$

$$\phi(0) = 0, \quad \phi'(0) = 0, \tag{A.20b}$$

$$\phi''(1) = 0, \quad -\phi'''(1) - 2\nu\phi'(1) = \alpha Q_B. \tag{A.20c}$$

[3] As usual, this notation understands that the real *or* the imaginary part of any complex expression must be taken.

They are of the type of Eqs. A.13, with the (static) stiffness operator $\mathcal{K} := \partial_s^4 + 2(\mu + \nu)\partial_s^2 + k_1$ being replaced by the dynamic stiffness operator $\mathcal{S}(\Omega) := \partial_s^4 + 2(\mu + \nu)\partial_s^2 + (k_1 + i\Omega\xi - \Omega^2)$.

Green Identity

To build up the Extended Green Identity, we follow the same steps of the previous section but, since we have to operate in complex field, we use the Hermitian scalar product, by multiplying the field Eq. A.20a by the complex conjugate of the dummy functions $\psi(s)$, denoted by $\bar{\psi}(s)$, thus obtaining:

$$\begin{aligned}
&\int_0^1 \bar{\psi}\left(\phi'''' + 2(\mu+\nu)\,\phi'' + \left(k_1 + i\Omega\xi - \Omega^2\right)\phi\right) \mathrm{d}s = \\
&\int_0^1 \phi\left(\bar{\psi}'''' + 2(\mu+\nu)\,\bar{\psi}'' + \left(k_1 + i\Omega\xi - \Omega^2\right)\bar{\psi}\right) \mathrm{d}s \\
&+\left[\bar{\psi}\phi''' - \bar{\psi}'\phi'' + \left(\bar{\psi}'' + 2(\mu+\nu)\,\bar{\psi}\right)\phi' - \left(\bar{\psi}''' + 2(\mu+\nu)\,\bar{\psi}'\right)\phi\right]_0^1.
\end{aligned} \tag{A.21}$$

From this, after conjugation of all terms, the adjoint homogeneous problem follows:

$$\psi'''' + 2(\mu+\nu)\psi'' + (k_1 - i\Omega\xi - \Omega^2)\psi = 0, \tag{A.22a}$$

$$\psi(0) = 0, \quad \psi'(0) = 0, \tag{A.22b}$$

$$\psi''(1) + 2\mu\psi(1) = 0, \quad -\psi'''(1) - 2(\mu+\nu)\psi'(1) = 0. \tag{A.22c}$$

Compatibility Condition

The compatibility condition reads:

$$\int_0^1 \bar{\psi}(s)\,q(s)\,\mathrm{d}s + \bar{\psi}(1)Q_B = 0, \quad \forall\,\psi(s) \in \mathcal{P}, \tag{A.23}$$

where $\mathcal{P}$ denotes the set of the solutions $\psi(s)$ of the adjoint homogeneous problem, Eq. A.22.

Normalization Condition

The solution of Eqs. A.20, in the resonant case, reads:

$$\phi(s) = \phi_p(s) + c_h\phi_h(s), \quad \forall c_h, \tag{A.24}$$

where ϕ_p is a particular solution, and $\phi_h(s)$ is the solution of the homogeneous problem, e.g., normalized as $\int_0^1 \phi_h^2(s)\mathrm{d}s = 1$. Normalization can be enforced as:

$$\mathcal{L}(\phi) := \int_0^1 w(s)\,\phi(s)\,\mathrm{d}s = A. \tag{A.25}$$

A.5 Exercises

The concepts illustrated in this chapter are applied to solve specific problems. Namely: (i) the adjoint problem for a beam on elastic soil is built-up, and the compatibility equation derived in presence of loads, as well as displacements prescribed at constraints; (ii) compatibility and normalization are then used to find the response of a critically compressed beam to assigned loads.

Exercise A.1 (*Adjoint problem and compatibility condition for a beam on elastic soil*) Consider the static problem for the system in Fig. A.1, where a roller, added at the end B of the sample beam, prescribes a vertical displacement $\breve{u}_B$, and, moreover, a couple C_B is applied at the same point. By using nondimensional quantities: (a) derive the adjoint problem and discuss adjointness; (b) derive the compatibility condition and interpret it on a mechanical ground.

(a) The boundary value problem reads:

$$u'''' + 2(\mu + \nu)u'' + k_1 u = p, \tag{A.26a}$$

$$u(0) = 0, \quad u'(0) = 0, \tag{A.26b}$$

$$u(1) = \breve{u}_B, \quad u''(1) = C_B. \tag{A.26c}$$

By substituting Eqs. A.26, *rendered homogeneous*, in the Extended Green Identity, Eq. A.14, and accounting for the arbitrariness of $u''(0)$, $u'''(0)$, $u'(1)$, $u'''(1)$, the adjoint homogeneous problem follows:

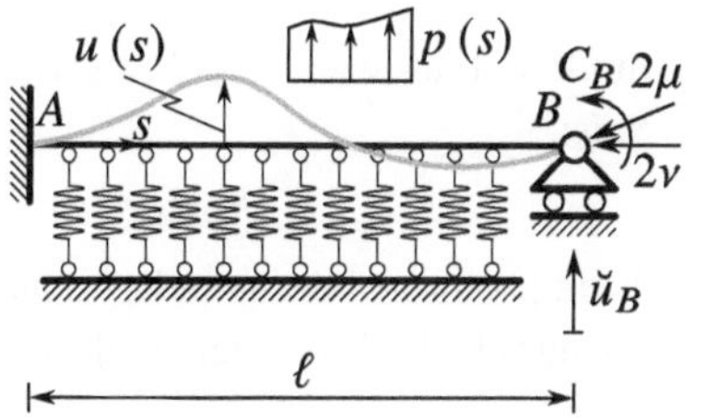

Fig. A.1 Linear beam resting on linear Winkler soil, under distributed force $p(s)$, tip dead force 2ν and follower force 2μ, tip moment C_B and prescribed vertical displacement $\breve{u}_B$

$$v'''' + 2(\mu + \nu)v'' + k_1 v = 0, \tag{A.27a}$$

$$v(0) = 0, \quad v'(0) = 0, \quad v(1) = 0, \quad v''(1) = 0. \tag{A.27b}$$

Since the two problems are identical, the original one is self-adjoint. This is due to the fact the roller absorbs the vertical component of the follower force, that therefore behaves as it were a dead load.

(b) By substituting Eqs. A.26 (now *non-homogeneous*) and Eqs. A.27 in the Extended Green Identity, the compatibility condition is derived:

$$\int_0^1 v(s)p(s)\mathrm{d}s + v'(1)C_B = -[v'''(1) + 2(\mu + \nu)v'(1)]\breve{u}_B, \quad \forall v(s) \in \mathcal{V}. \tag{A.28}$$

It expresses the following statement: the virtual work spent by the true active forces p, C_B, in the associated v-displacements, equates the virtual work spent by the adjoint reactive vertical force at B, namely $R_B := -[v'''(1) + 2(\mu + \nu)v'(1)]$, in the prescribed true displacement $\breve{u}_B$. □

Exercise A.2 (*Static response of a critically compressed beam to assigned loads*) Consider the static problem for the beam of Fig. A.1 when the elastic soil is removed ($k_1 = 0$), under the action of a transverse uniform force $p(s) = \bar{p}$ and a compression dead force 2ν, while $C_B = 0$ and $\mu = 0$. (a) Evaluate the critical load ν_c. (b) Compute the vertical displacement $\breve{u}_B$, prescribed by the roller, able to guarantee the solvability of the problem. (c) Derive the static response of the beam, by using $u'(1) = 1$ as normalization condition.

(a) The boundary value problem is:

$$u'''' + 2\nu u'' = \bar{p}, \tag{A.29a}$$

$$u(0) = 0, \quad u'(0) = 0, \tag{A.29b}$$

$$u(1) = \breve{u}_B, \quad u''(1) = 0. \tag{A.29c}$$

Equations A.29a, made homogeneous, admits the following characteristic equation:

$$\lambda^4 + 2\nu\lambda^2 = 0, \tag{A.30}$$

whose roots are $\lambda_{1,2} = \pm i\beta$, and $\lambda_{3,4} = 0$, where $\beta := \sqrt{2\nu}$. Therefore, the general solution reads:

$$u_c(s) = a_1 \cos(\beta s) + a_2 \sin(\beta s) + a_3 s + a_4. \tag{A.31}$$

The imposition of the boundary conditions, Eqs. A.29b, c, with $\breve{u}_B = 0$, gives rise to a homogeneous algebraic problem in the unknowns a_j, i.e., $\mathbf{Aa} = \mathbf{0}$, where $a = (a_1\ a_2\ a_3\ a_4)^T$ and

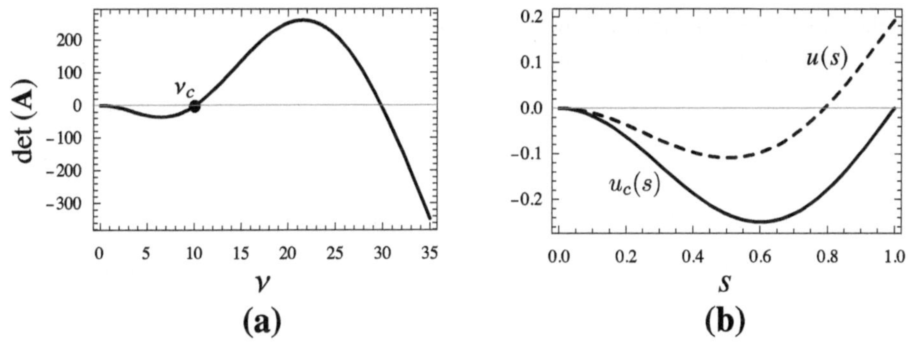

Fig. A.2 Solution of the boundary value problem in Eqs. A.29: **a** determinant of the matrix **A** *versus* ν; **b** critical mode $u_c(s)$ (continuous line) and static response $u(s)$, when $\bar{p} = 5$ (dashed line)

$$\mathbf{A} := \begin{pmatrix} 1 & 0 & 0 & 1 \\ 0 & \beta & 1 & 0 \\ \cos\beta & \sin\beta & 1 & 1 \\ -\beta^2\cos\beta & -\beta^2\sin\beta & 0 & 0 \end{pmatrix}. \tag{A.32}$$

The critical load ν_c is the lower value of ν which makes singular the matrix **A**. In particular, $\det \mathbf{A} = \beta^2(\beta\cos\beta - \sin\beta)$ so that, excluding the trivial solutions $\beta = 0$, the values of β vanishing $\det \mathbf{A}$ are the roots of

$$\tan\beta = \beta, \tag{A.33}$$

which can be numerically solved. The plot of $\det \mathbf{A}$ *versus* ν is shown in Fig. A.2a. It turns out the $\nu_c = 10.095$ (corresponding to $\beta = 4.493$), which provides the solutions $\mathbf{a} = (0.573a\ -0.127a\ 0.573a\ -0.573a)^T$, $\forall a$; the corresponding critical mode $u_c(s)$ is shown in Fig. A.2b, where the normalization condition $u_c''(1) = 1$ is used (i.e., $a = 0.312$).

(b) Since the boundary value problem Eqs. A.29 is self-adjoint, it turns out that the left eigenvector coincides with $u_c(s)$. Therefore, the solvability condition becomes:

$$\int_0^1 \bar{p}u_c(s)\mathrm{d}s - R_B\breve{u}_B = 0, \tag{A.34}$$

where $R_B := -(u_c'''(1) + 2\nu_c u_c'(1))$, which represents the reactive vertical force in correspondence of the roller at B, in the critical condition.[4] In particular, from Eq. A.34, it turns out that:

$$\breve{u}_B = \frac{\bar{p}\int_0^1 u_c(s)\mathrm{d}s}{R_B} = 0.0385\bar{p} \tag{A.35}$$

[4] See the Exercise A.1 for the physical interpretation of Eq. A.34.

is the only value of $\breve{u}_B$ which allows to solve Eqs. A.29.[5]

(c) Equation A.29a has solution:

$$u(s) = u_h(s) + u_p(s), \tag{A.36}$$

where:

$$u_h(s) = \tilde{a}_1 \cos(\beta s) + \tilde{a}_2 \sin(\beta s) + \tilde{a}_3 s + \tilde{a}_4 \tag{A.37}$$

is the complementary solution, in which $\tilde{a}_j$ ($j = 1, \ldots, 4$) are new arbitrary constants (different from a_j), and

$$u_p(s) = \frac{\bar{p}s^2}{2\beta^2} \tag{A.38}$$

is the particular solution (easily evaluated, due to the uniformity of $\bar{p}$[6]). Coefficients $\tilde{a}_j$ in Eq. A.36 must be computed by the boundary conditions, Eqs. A.29b, c. However, due to the singularity of the problem and to the holding of compatibility, only three conditions among them are independent, providing a solution which is indeterminate. Therefore one of the geometric boundary conditions, e.g., the first one in Eq. A.29b, can be replaced by the normalization condition, here taken as $u'(1) = 1$. The resulting function $u(s)$ has expression:

$$\begin{aligned} u(s) =& (0.178 - 0.011\,\bar{p}) \cos(4.49341\, s) - 0.040 \sin(4.49341\, s) + 0.178\, s \\ &+ (0.011\bar{p} - 0.178) + 0.025\bar{p}s^2, \end{aligned} \tag{A.39}$$

represented in Fig. A.2b for $\bar{p} = 5$. □

Reference

1. Lanczos, C.: Linear differential operators. Society for Industrial and Applied Mathematics. Philadelphia (1996)

[5] Note that the normalization factor does not count in Eq. A.35, because it is present both at numerator and denominator.

[6] See Exercise 2.1 for details on the evaluation of the particular solutions in more complex cases.

Discrete Approach: The Galerkin Method

B

B.1 Galerkin Discretization

The problems tackled in the book are relevant to continuous systems, governed by partial differential equations and boundary conditions. However, we know from Computational Mechanics that it is possible to transform the equations into time ordinary differential equations by using space discretization through the *Galerkin* or *Finite Element* methods. Here we confine ourselves to the Galerkin approach, aimed to transform the original time-space partial differential equations in ordinary time-differential equations [1, 2].

B.1.1 Discrete Model of Beam on Winkler Soil

Let us consider again Eqs. 1.7, governing the motion of the beam on elastic soil, and let us put them in a *weak form* by performing the scalar product with a kinematically admissible *test function* $\delta u(s,t)$, representing a virtual displacement:

$$\begin{aligned}&\int_0^1\left[\ddot{u}+\xi\dot{u}+u''''+2(\mu+\nu)u''+\sum_{i=1}^{\infty}k_i u^i-\alpha p(s,t)\right]\delta u\,\mathrm{d}s\\&\quad+u''(1,t)\delta u'(1,t)+[-u'''(1,t)-2\nu u'(1,t)-\alpha P_B(t)]\delta u(1,t)=0,\end{aligned} \tag{B.1}$$

where $\nu := \nu_s + \nu_d \cos(\Omega t)$. Here, the integral term represents the virtual work expended by the unbalanced forces in the domain (the *residual* of the field equation) over the test displacement $\delta u(s,t)$, and the last two terms the unbalanced couple and force at the free end (the residuals at the boundary) over the relevant test rotation $\delta u'(1,t)$ and displacement $\delta u(1,t)$, respectively. Therefore, Eq. B.1 is the *Virtual Work Equation*. It must hold under the condition that the *geometrical* boundary conditions in Eq. 1.7b are satisfied.

A. Luongo et al., *Perturbation Methods and Nonlinear Phenomena*,
Synthesis Lectures on Engineering, Science, and Technology,
https://doi.org/10.1007/978-3-031-49397-3

According to the Galerkin Method, we express the unknown variable $u(s,t)$ as a linear combination of known function of space $\phi_k(s)$, said *trial functions*, and unknown function of time $x_k(t)$, said *Lagrangian coordinates*, i.e.:

$$u(s,t) = \sum_{k=1}^{N} \phi_k(s) x_k(t) \equiv \boldsymbol{\phi}(s)\mathbf{x}(t), \tag{B.2}$$

where $\boldsymbol{\phi} := (\phi_1(s) \ \dots \ \phi_N(s))$ is a row matrix and $\mathbf{x} := (x_1(t) \ \dots \ x_N(s))^T$ a column matrix. In order for $u(s,t)$ to satisfy the geometrical boundary conditions for any $x_k(t)$, *each* trial function is required to satisfy them separately, i.e., $\phi_k(0) = 0$, $\phi_k'(0) = 0$. Moreover, it is convenient (but not mandatory) to take the test functions in the same basis of the trial functions, i.e.:

$$\delta u(s,t) = \sum_{k=1}^{N} \phi_k(s)\, \delta x_k(t) \equiv \boldsymbol{\phi}(s)\, \delta\mathbf{x}(t). \tag{B.3}$$

By substituting the Eqs. B.2 and B.3 into the Virtual Work Eq. B.1, performing integration, and requiring that it holds for any $\delta\mathbf{x}$, a set of N ordinary differential equations in the $\mathbf{x}(t)$ unknown is derived, namely:

$$\mathbf{M}\ddot{\mathbf{x}} + \xi\mathbf{C}\dot{\mathbf{x}} + (\mathbf{K}_0 - 2\mu\mathbf{K}_\mu - 2\nu\mathbf{K}_\nu)\mathbf{x} + \mathbf{n}_2(\mathbf{x},\mathbf{x}) + \mathbf{n}_3(\mathbf{x},\mathbf{x},\mathbf{x}) + \cdots = \alpha\mathbf{p}, \tag{B.4}$$

Here, the following matrices and vectors have been introduced[7]:

$$\begin{aligned}
\mathbf{M} = \mathbf{C} &:= \int_0^1 \boldsymbol{\phi}^T(s)\,\boldsymbol{\phi}(s)\,\mathrm{d}s, \\
\mathbf{K}_0 &:= \int_0^1 \boldsymbol{\phi}''^T(s)\boldsymbol{\phi}''(s)\,\mathrm{d}s + k_1 \int_0^1 \boldsymbol{\phi}^T(s)\boldsymbol{\phi}(s)\mathrm{d}s, \\
\mathbf{K}_\mu &:= \int_0^1 \boldsymbol{\phi}'^T(s)\boldsymbol{\phi}'(s)\,\mathrm{d}s - \boldsymbol{\phi}^T(1)\boldsymbol{\phi}'(1), \\
\mathbf{K}_\nu &:= \int_0^1 \boldsymbol{\phi}'^T(s)\boldsymbol{\phi}'(s)\,\mathrm{d}s, \\
\mathbf{p} &:= \int_0^1 \boldsymbol{\phi}^T(s)p(s,t)\,\mathrm{d}s + \boldsymbol{\phi}^T(1)P_B(t), \\
\mathbf{n}_2(\mathbf{x},\mathbf{y}) &:= k_2 \left\{ \sum_{j=1}^{N}\sum_{k=1}^{N} n_{ijk}x_j y_k \right\}, \\
\mathbf{n}_3(\mathbf{x},\mathbf{y},\mathbf{z}) &:= k_3 \left\{ \sum_{j=1}^{N}\sum_{k=1}^{N}\sum_{l=1}^{N} n_{ijkl}x_j y_k z_l \right\},
\end{aligned} \tag{B.5}$$

[7] Integration by parts have been performed to simplify the expression of $\mathbf{K}_0$, $\mathbf{K}_\mu$ and $\mathbf{K}_\nu$.

where[8]:

$$n_{ijk} := \int_0^1 \phi_i(s)\,\phi_j(s)\,\phi_k(s)\,\mathrm{d}s, \qquad n_{ijkl} := \int_0^1 \phi_i(s)\,\phi_j(s)\,\phi_k(s)\,\phi_l(s)\,\mathrm{d}s, \tag{B.6}$$

In Eqs. B.4 and B.5, capital and lower-case bold symbols denote $N \times N$ matrices and N-column matrices, respectively: $\mathbf{M} = \mathbf{M}^T$ is the *mass* matrix, $\mathbf{C} = \mathbf{C}^T$ the *damping* matrix, $\mathbf{K}_0 = \mathbf{K}_0^T$ the *elastic stiffness* matrix, $\mathbf{K}_\mu \neq \mathbf{K}_\mu^T$ and $\mathbf{K}_\nu = \mathbf{K}_\nu^T$ the *geometric stiffness* matrices, associated with the nonconservative and conservative (or time-dependent) prestress forces, respectively. The non-symmetry of $\mathbf{K}_\mu$ accounts for *non-self-adjointness* of the problem. Moreover, $\mathbf{p}(t)$ is the time-depending *load vector* and $\mathbf{n}_2(\mathbf{x}, \mathbf{x})$, $\mathbf{n}_3(\mathbf{x}, \mathbf{x}, \mathbf{x})$ are the *nonlinear force vectors*, generated by the nonlinear part of the elastic forces.[9]

Remark B.1 An important task in using the Galerkin method concerns the choice of the trial functions. This selection, of course, influences the approximation we can obtain when apply the perturbation method to the space-discretized system, instead of the continuous system. In order for the solutions of the two problems to be close each other, the basis of the trial function should be able to reproduce the generating (active coordinates) and higher-order (passive coordinates) solutions of the continuous perturbation equations. We observed that these latter (denoted by $\chi(s)$) describe the modification of the space-dependence of the solution with respect to the linear problem. To within the truncation adopted, *this modification is absent in symmetric systems*, since it just depends on quadratic nonlinearities. Therefore, for symmetric systems, if we adopt as *unique* ($N = 1$) trial function the solution of the linear

[8] Note that $\mathbf{n}_2$ and $\mathbf{n}_3$ are column matrices, where i is the index of the row ($i = 1, \ldots, N$).

[9] While the reader is certainly accustomed with matrices and vectors, he could be novice with the symbolism adopted for nonlinear terms. Therefore, some additional words are spent about it. It is convenient, for further developments, to think $\mathbf{n}_2(\mathbf{x}, \mathbf{x})$ and $\mathbf{n}_3(\mathbf{x}, \mathbf{x}, \mathbf{x})$ as *functions* of their argument, namely as operators which work on two or three input-vectors, and transform them into just one output-vector. Moreover, it is suitable to extend their definition to the case in which their arguments differ among them (e.g., $\mathbf{x}, \mathbf{y}, \mathbf{z}$), which represent three different sets of values assumed by the variable $\mathbf{x}$; for this reason we have written $\mathbf{n}_2(\mathbf{x}, \mathbf{y})$ and $\mathbf{n}_3(\mathbf{x}, \mathbf{y}, \mathbf{z})$ in the defining Eqs. B.5. Since these functions are multilinear (namely bi- or three-linear) forms of their arguments, with symmetric coefficients, the following properties hold:

$$\begin{aligned}
P1: &\quad \mathbf{n}_2(\mathbf{x}, \mathbf{y}) = \mathbf{n}_2(\mathbf{y}, \mathbf{x}), \quad \mathbf{n}_3(\mathbf{x}, \mathbf{y}, \mathbf{z}) = \mathbf{n}_3(\mathbf{x}, \mathbf{z}, \mathbf{y}) = \mathbf{n}_3(\mathbf{y}, \mathbf{z}, \mathbf{x}) = \cdots, \\
P2: &\quad \mathbf{n}_2(\alpha\mathbf{x}, \beta\mathbf{y}) = \alpha\beta\mathbf{n}_2(\mathbf{x}, \mathbf{y}), \quad \mathbf{n}_3(\alpha\mathbf{x}, \beta\mathbf{y}, \gamma\mathbf{z}) = \alpha\beta\gamma\mathbf{n}_3(\mathbf{x}, \mathbf{y}, \mathbf{z}), \\
P3: &\quad \mathbf{n}_2(\mathbf{x} + \mathbf{y}, \mathbf{x} + \mathbf{y}) = \mathbf{n}_2(\mathbf{x}, \mathbf{x}) + 2\mathbf{n}_2(\mathbf{x}, \mathbf{y}) + \mathbf{n}_2(\mathbf{y}, \mathbf{y}),
\end{aligned}$$

where α, β, γ are arbitrary scalars. Property $P1$ expresses insensitivity to arguments permutation, consequent to symmetry of the coefficients n_{ijk} and n_{ijkl} with respect to their indexes; properties $P1$, $P2$ can be easily checked by using definitions of the nonlinear vectors. Remarkably, in performing the operations indicated above, the nonlinear vectors can be handled as they were scalar quantities. All these properties will be used ahead.

problem, we obtain a single-d.o.f. system whose first-order perturbation solution coincides with that of the continuous system!

B.1.2 Discrete Metamodel

The Galerkin method is now applied to the metamodel defined by Eqs. 5.1. By splitting the boundary conditions in *geometric* (holding on the portion Γ_u of Γ) and *mechanic* (holding on the supplementary part Γ_f), and introducing a load multiplier α, the equations read:

$$\mathcal{M}\ddot{\mathbf{u}} + \mathcal{C}\dot{\mathbf{u}} + \mathcal{K}\mathbf{u} + \boldsymbol{n}(\mathbf{u}) = \alpha\mathbf{b}, \quad \text{in } \Omega, \tag{B.7a}$$

$$\mathcal{B}_u\mathbf{u} = \mathbf{0}, \quad \text{on } \Gamma_u, \tag{B.7b}$$

$$\mathcal{D}\dot{\mathbf{u}} + \mathcal{B}_f\mathbf{u} = \alpha\mathbf{f}, \quad \text{on } \Gamma_f, \tag{B.7c}$$

where $\mathbf{u} = \mathbf{u}(\mathbf{s}, t)$, $\mathbf{b} = \mathbf{b}(\mathbf{s}, t)$, $\mathbf{f} = \mathbf{f}(\mathbf{s}, t)$; moreover, $\boldsymbol{n}(\mathbf{u}) := \boldsymbol{n}_2(\mathbf{u}, \mathbf{u}) + \boldsymbol{n}_3(\mathbf{u}, \mathbf{u}, \mathbf{u}) + \cdots$. The relevant Virtual Work Equation is:

$$\int_0^1 \delta\mathbf{u}^T[\mathcal{M}\ddot{\mathbf{u}} + \mathcal{C}\dot{\mathbf{u}} + \mathcal{K}\mathbf{u} + \boldsymbol{n}(\mathbf{u}) - \alpha\mathbf{b}]\mathrm{d}\Omega + \int_{\Gamma_f} \delta\mathbf{u}^T[\mathcal{D}\dot{\mathbf{u}} + \mathcal{B}_f\mathbf{u} - \alpha\mathbf{f}]\mathrm{d}\Gamma = 0, \tag{B.8}$$

in which $\delta\mathbf{u}(\mathbf{s}, t)$ is a test function. The displacement field is expressed in the form:

$$\mathbf{u}(\mathbf{s}, t) = \boldsymbol{\Phi}(\mathbf{s})\,\mathbf{x}(t), \tag{B.9}$$

where $\boldsymbol{\Phi}(\mathbf{x})$ is a $M \times N$ matrix of known functions of the spatial coordinates (M being the number of the components of $\mathbf{u}$), satisfying the geometric boundary conditions on Γ_u; $\mathbf{x}(t)$ is a column matrix of unknown Lagrangian parameters. With Eqs. B.9, B.8 becomes:

$$\begin{aligned} &\int_\Omega \delta\mathbf{x}^T\boldsymbol{\Phi}^T\left[\mathcal{M}\boldsymbol{\Phi}\ddot{\mathbf{x}} + \mathcal{C}\boldsymbol{\Phi}\dot{\mathbf{x}} + \mathcal{K}\boldsymbol{\Phi}\mathbf{x} + \boldsymbol{n}(\boldsymbol{\Phi}\mathbf{x}) - \alpha\mathbf{b}\right]\mathrm{d}\Omega \\ &+ \int_{\Gamma_f} \delta\mathbf{x}^T\boldsymbol{\Phi}^T\left[\mathcal{D}\boldsymbol{\Phi}\dot{\mathbf{x}} + \mathcal{B}_f\boldsymbol{\Phi}\mathbf{x} - \alpha\mathbf{f}\right]\mathrm{d}\Gamma = 0. \end{aligned} \tag{B.10}$$

By requiring it holds for any $\delta\mathbf{x}$, it follows:

$$\mathbf{M}\ddot{\mathbf{x}} + \mathbf{C}\dot{\mathbf{x}} + \mathbf{K}\mathbf{x} + \mathbf{n}_2(\mathbf{x}, \mathbf{x}) + \mathbf{n}_3(\mathbf{x}, \mathbf{x}, \mathbf{x}) + \cdots = \alpha\mathbf{p}, \tag{B.11}$$

where:

$$
\begin{aligned}
\mathbf{M} &:= \int_\Omega \boldsymbol{\Phi}^T(\mathbf{s})\mathcal{M}\boldsymbol{\Phi}(\mathbf{s})\,\mathrm{d}\Omega, \\
\mathbf{C} &:= \int_\Omega \boldsymbol{\Phi}^T(\mathbf{s})\mathcal{C}\boldsymbol{\Phi}(\mathbf{s})\,\mathrm{d}\Omega + \int_{\Gamma_f} \boldsymbol{\Phi}^T\mathcal{D}\boldsymbol{\Phi}\,\mathrm{d}\Gamma, \\
\mathbf{K} &:= \int_\Omega \boldsymbol{\Phi}^T(\mathbf{s})\mathcal{K}\boldsymbol{\Phi}(\mathbf{s})\,\mathrm{d}\Omega + \int_{\Gamma_f} \boldsymbol{\Phi}^T\mathcal{B}_f\boldsymbol{\Phi}\,\mathrm{d}\Gamma, \\
\mathbf{p} &:= \int_\Omega \boldsymbol{\Phi}^T(\mathbf{s})\mathbf{b}(\mathbf{s},t)\,\mathrm{d}\Omega + \int_{\Gamma_f} \boldsymbol{\Phi}^T(\mathbf{s})\mathbf{f}(\mathbf{s},t)\,\mathrm{d}\Gamma.
\end{aligned}
\tag{B.12}
$$

Moreover, the nonlinear vectors are drawn from the following identities:

$$
\begin{aligned}
\mathbf{n}_2(\mathbf{x},\mathbf{y}) &:= \int_\Omega \boldsymbol{\Phi}^T n_2\left(\boldsymbol{\Phi}\mathbf{x},\boldsymbol{\Phi}\mathbf{y}\right)\,\mathrm{d}\Omega, \\
\mathbf{n}_3(\mathbf{x},\mathbf{y},\mathbf{z}) &:= \int_\Omega \boldsymbol{\Phi}^T n_3\left(\boldsymbol{\Phi}\mathbf{x},\boldsymbol{\Phi}\mathbf{y},\boldsymbol{\Phi}\mathbf{z}\right)\,\mathrm{d}\Omega.
\end{aligned}
\tag{B.13}
$$

Reference will be made ahead to the sample beam only, and therefore to the discretized Eq. B.4. However, the extension to the metamodel, or any other models, is straightforward.

B.2 Nonlinear Statics

When the beam is unprestressed and loaded by transverse loads only, equilibrium is governed by algebraic equations deduced by Eqs. B.4, $\mathbf{x}$ being independent of time, namely:

$$
\mathbf{K}_0\mathbf{x} + \mathbf{n}_2(\mathbf{x},\mathbf{x}) + \mathbf{n}_3(\mathbf{x},\mathbf{x},\mathbf{x}) + \cdots = \alpha\mathbf{p}. \tag{B.14}
$$

As we did for the continuous problem (Sect. 2.2), we use the rescaling $\mathbf{x} \to \epsilon\hat{\mathbf{x}}$, $\alpha \to \epsilon\hat{\alpha}$, and expand the unknown as:

$$
\mathbf{x} = \mathbf{x}_0 + \epsilon\mathbf{x}_1 + \epsilon^2\mathbf{x}_2 + \cdots. \tag{B.15}
$$

By virtue of the properties of the nonlinear vectors (see Note 3), it is $\mathbf{n}_2(\mathbf{x},\mathbf{x}) = \mathbf{n}_2(\mathbf{x}_0,\mathbf{x}_0) + 2\epsilon\mathbf{n}_2(\mathbf{x}_0,\mathbf{x}_1) + \cdots$, and $\mathbf{n}_3(\mathbf{x},\mathbf{x},\mathbf{x}) = \mathbf{n}_3(\mathbf{x}_0,\mathbf{x}_0,\mathbf{x}_0) + \cdots$, so that the perturbation equations are:

$$
\epsilon^0 : \mathbf{K}_0\mathbf{x}_0 = \alpha\mathbf{p}, \tag{B.16a}
$$

$$
\epsilon^1 : \mathbf{K}_0\mathbf{x}_1 = -\mathbf{n}_2(\mathbf{x}_0,\mathbf{x}_0), \tag{B.16b}
$$

$$
\epsilon^2 : \mathbf{K}_0\mathbf{x}_2 = -2\mathbf{n}_2(\mathbf{x}_0,\mathbf{x}_1) - \mathbf{n}_3(\mathbf{x}_0,\mathbf{x}_0,\mathbf{x}_0). \tag{B.16c}
$$

By solving them in sequence, and reabsorbing the perturbation parameter, we find:

$$\mathbf{x} = \alpha \mathbf{w}_0 + \alpha^2 \mathbf{w}_1 + \alpha^3 \mathbf{w}_2 + \cdots, \tag{B.17}$$

where the vectors $\mathbf{w}_j$ are solutions of the linear algebraic nonsingular problems:

$$\epsilon^0 : \mathbf{K}_0 \mathbf{w}_0 = \mathbf{p}, \tag{B.18a}$$
$$\epsilon^1 : \mathbf{K}_0 \mathbf{w}_1 = -\mathbf{n}_2(\mathbf{w}_0, \mathbf{w}_0), \tag{B.18b}$$
$$\epsilon^2 : \mathbf{K}_0 \mathbf{w}_2 = -2\mathbf{n}_2(\mathbf{w}_0, \mathbf{w}_1) - \mathbf{n}_3(\mathbf{w}_0, \mathbf{w}_0, \mathbf{w}_0), \tag{B.18c}$$

which are the discrete version of Eqs. 2.10 and 2.9, respectively.

B.3 Buckling and Postbuckling of Static Systems

When the beam is prestressed by a dead load, and no transverse loads act on it, equilibrium is governed by Eqs. B.4, leading to the following algebraic equations[10]:

$$(\mathbf{K}_0 - 2\nu \mathbf{K}_\nu)\mathbf{x} + \mathbf{n}_2(\mathbf{x}, \mathbf{x}) + \mathbf{n}_3(\mathbf{x}, \mathbf{x}, \mathbf{x}) + \cdots = \mathbf{0}. \tag{B.19}$$

To analyze buckling, as we did for the continuous problem (Sect. 2.3), we rescale the variables as $\mathbf{x} \to \epsilon \hat{\mathbf{x}}$ and expand them together with the bifurcation parameter as:

$$\mathbf{x} = \mathbf{x}_0 + \epsilon \mathbf{x}_1 + \epsilon^2 \mathbf{x}_2 + \cdots, \tag{B.20a}$$
$$\nu = \nu_0 + \epsilon \nu_1 + \epsilon^2 \nu_2 + \cdots, \tag{B.20b}$$

from which the perturbation equations follow:

$$\epsilon^0 : (\mathbf{K}_0 - 2\nu_0 \mathbf{K}_\nu)\mathbf{x}_0 = \mathbf{0}, \tag{B.21a}$$
$$\epsilon^1 : (\mathbf{K}_0 - 2\nu_0 \mathbf{K}_\nu)\mathbf{x}_1 = 2\nu_1 \mathbf{K}_\nu \mathbf{x}_0 - \mathbf{n}_2(\mathbf{x}_0, \mathbf{x}_0), \tag{B.21b}$$
$$\epsilon^2 : (\mathbf{K}_0 - 2\nu_0 \mathbf{K}_\nu)\mathbf{x}_2 = 2\nu_2 \mathbf{K}_\nu \mathbf{x}_0 + 2\nu_1 \mathbf{K}_\nu \mathbf{x}_1 - 2\mathbf{n}_2(\mathbf{x}_0, \mathbf{x}_1) - \mathbf{n}_3(\mathbf{x}_0, \mathbf{x}_0, \mathbf{x}_0). \tag{B.21c}$$

Equation B.21a is an algebraic eigenvalue problem in the bifurcation parameter ν_0. The smallest eigenvalue $\nu_0 = \nu_c$ is the critical load; the associated (real) eigenvector $\mathbf{u}_c$ is the critical mode, i.e., $(\mathbf{K}_0 - 2\nu_c \mathbf{K}_\nu)\mathbf{u}_c = \mathbf{0}$ is satisfied; hence, the solution of Eq. B.21a is $\mathbf{x}_0 = a\mathbf{u}_c$, with a the undetermined amplitude and $\mathbf{u}_c$ normalized as desired (e.g., $\mathbf{u}_c^T \mathbf{u}_c = 1$, or $\mathbf{e}_h^T \mathbf{u}_c = 1$, with $\mathbf{e}_h$ the hth canonical vector).

In order to satisfy the perturbation Eq. B.21b, the known term on its right hand side must be rendered orthogonal to the left eigenvector $\mathbf{v}_c$, which solves the problem

[10] Index s dropped on ν.

$(\mathbf{K}_0 - 2\nu_c \mathbf{K}_\nu)^T \mathbf{v}_c = \mathbf{0}$; however, since the eigenvalue problem is self-adjoint, it is $\mathbf{v}_c = \mathbf{u}_c$, and therefore compatibility leads to $\nu_1 = C_1 a$, with:

$$C_1 := \frac{1}{2} \frac{\mathbf{u}_c^T \mathbf{n}_2(\mathbf{u}_c, \mathbf{u}_c)}{\mathbf{u}_c^T \mathbf{K}_\nu \mathbf{u}_c}. \tag{B.22}$$

Solution of Eq. B.21b then reads $\mathbf{x}_1 = a^2 \mathbf{w}_1$, with $\mathbf{w}_1$ solution of the *singular* problem:

$$(\mathbf{K}_0 - 2\nu_c \mathbf{K}_\nu)\mathbf{w}_1 = 2C_1 \mathbf{K}_\nu \mathbf{u}_c - \mathbf{n}_2(\mathbf{u}_c, \mathbf{u}_c), \tag{B.23}$$

made unique by a suitable normalization condition (e.g., $\mathbf{u}_c^T \mathbf{w}_1 = 0$, or $\mathbf{e}_h^T \mathbf{w}_1 = 0$). From compatibility of the next perturbation equation, $\nu_2 = C_2 a^2$ is finally obtained, with:

$$C_2 := \frac{1}{2} \frac{2\mathbf{u}_c^T \mathbf{n}_2(\mathbf{u}_c, \mathbf{w}_1) + \mathbf{u}_c^T \mathbf{n}_3(\mathbf{u}_c, \mathbf{u}_c, \mathbf{u}_c) - 2C_1 \mathbf{u}_c^T \mathbf{K}_\nu \mathbf{w}_1}{\mathbf{u}_c^T \mathbf{K}_\nu \mathbf{u}_c}. \tag{B.24}$$

By summarizing, after reabsorbing the perturbation parameter, the solution reads:

$$\mathbf{x} = a\mathbf{u}_c + a^2 \mathbf{w}_1 + \cdots, \tag{B.25a}$$

$$\nu = \nu_c + C_1 a + C_2 a^2 + \cdots, \tag{B.25b}$$

which is the discrete counterpart of Eqs. 2.34.

B.4 External Resonances

When the beam is unprestressed and subjected to harmonic transverse loads, namely $\mathbf{p}(t) = \mathbf{q} \cos \Omega t$, the equations of motion, Eqs. B.4, read:

$$\mathbf{M}\ddot{\mathbf{x}} + \xi \mathbf{C}\dot{\mathbf{x}} + \mathbf{K}_0 \mathbf{x} + \mathbf{n}_2(\mathbf{x}, \mathbf{x}) + \mathbf{n}_3(\mathbf{x}, \mathbf{x}, \mathbf{x}) + \cdots = \alpha \mathbf{q} \cos \Omega t. \tag{B.26}$$

The primary resonance case, the sub- and super-harmonic resonances are separately dealt with.

B.4.1 Primary Resonance

We consider a primary resonant excitation (and no internal resonance condition), by letting $\Omega = \omega_r + \sigma_r$, where σ_r is a small detuning and ω_r is the rth natural frequency of the *discretized* undamped beam. Therefore, frequency ω_r solves the eigenvalue problem $(\mathbf{K}_0 - \omega_r^2 \mathbf{M})\mathbf{u}_r = \mathbf{0}$, with $\mathbf{u}_r$ the associated (real) natural mode, suitably normalized. Due to self-adjointness ($\mathbf{K}_0$ and $\mathbf{M}$ are both symmetric), $\mathbf{v}_r = \mathbf{u}_r$ holds. By proceeding in the same way followed for the continuous model (Sect. 3.2), we rescale the dependent variable and the small parameters as $\mathbf{x} \to \epsilon\hat{\mathbf{x}}$, $\alpha \to \epsilon^3\hat{\alpha}$, $\sigma_r \to \epsilon^2\hat{\sigma}_r$, $\xi \to \epsilon^2\hat{\xi}$. Moreover, we expand the variable as $\mathbf{x} = \mathbf{x}_0 + \epsilon\mathbf{x}_1 + \epsilon^2\mathbf{x}_2 + \cdots$ and, according to the Multiple Scale Method (MSM), the time-differential operators as $\frac{\mathrm{d}}{\mathrm{d}t} = \mathrm{d}_0 + \epsilon\mathrm{d}_1 + \epsilon^2\mathrm{d}_2 + \cdots$, $\frac{\mathrm{d}^2}{\mathrm{d}t^2} = \mathrm{d}_0^2 + 2\epsilon\mathrm{d}_0\mathrm{d}_1 + \epsilon^2\left(\mathrm{d}_1^2 + 2\mathrm{d}_0\mathrm{d}_2\right) + \cdots$, where $\mathrm{d}_k := \partial/\partial t_k$ denotes partial derivative with respect $t_k := \epsilon^k t$, $k = 0, 1, \ldots$.[11] The following perturbation equations are derived:

$$
\begin{aligned}
\epsilon^0 &: \mathbf{M}\mathrm{d}_0^2\mathbf{x}_0 + \mathbf{K}_0\mathbf{x}_0 = \mathbf{0}, && \text{(B.27a)}\\
\epsilon^1 &: \mathbf{M}\mathrm{d}_0^2\mathbf{x}_1 + \mathbf{K}_0\mathbf{x}_1 = -2\mathbf{M}\mathrm{d}_0\mathrm{d}_1\mathbf{x}_0 - \mathbf{n}_2(\mathbf{x}_0, \mathbf{x}_0), && \text{(B.27b)}\\
\epsilon^2 &: \mathbf{M}\mathrm{d}_0^2\mathbf{x}_2 + \mathbf{K}_0\mathbf{x}_2 = -\mathbf{M}\mathrm{d}_1^2\mathbf{x}_0 - 2\mathbf{M}\mathrm{d}_0\mathrm{d}_2\mathbf{x}_0 - 2\mathbf{M}\mathrm{d}_0\mathrm{d}_1\mathbf{x}_1 && \text{(B.27c)}\\
&\quad - \xi\mathbf{C}\mathrm{d}_0\mathbf{x}_0 - 2\mathbf{n}_2(\mathbf{x}_0, \mathbf{x}_1) - \mathbf{n}_3(\mathbf{x}_0, \mathbf{x}_0, \mathbf{x}_0) + \left(\frac{1}{2}\alpha\mathbf{q}e^{i\sigma_r t_2}e^{i\omega_r t_0} + \text{c.c.}\right).
\end{aligned}
$$

The monomodal generating solution is $\mathbf{x}_0 = A_r(t_1, t_2, \ldots)\mathbf{u}_r e^{i\omega_r t_0} + \text{c.c.}$, with $A_r = \frac{1}{2}a_r(t_1, t_2, \ldots)e^{i\varphi_r(t_1, t_2, \ldots)}$ the complex amplitude. By using the properties in Note 3, it follows that $\mathbf{n}_2(\mathbf{x}_0, \mathbf{x}_0) = \left(A_r^2 e^{2i\omega_r t_0} + A_r\bar{A}_r\right)\mathbf{n}_2(\mathbf{u}_r, \mathbf{u}_r) + \text{c.c.}$, so that compatibility of Eq. B.27b, which requires orthogonality of the ω_r-harmonic excitation vector to $\mathbf{u}_r$, entails that $\mathrm{d}_1 A_r = 0$. By omitting, as usual, the complementary solution, Eq. B.27b then furnishes:

$$
\mathbf{x}_1 = A_r^2\mathbf{w}_{12}e^{2i\omega_r t_0} + A_r\bar{A}_r\mathbf{w}_{10} + \text{c.c.}, \tag{B.28}
$$

where $\mathbf{w}_{1m}$ $(m = 2, 0)$ are solutions of the *nonsingular* algebraic problems:

$$
(\mathbf{K}_0 - m^2\omega_r^2\mathbf{M})\mathbf{w}_{1m} = -\mathbf{n}_2(\mathbf{u}_r, \mathbf{u}_r). \tag{B.29}
$$

By going on to the next order, it turns out that:

$$
\begin{aligned}
\mathbf{n}_2(\mathbf{x}_0, \mathbf{x}_1) &= A_r^2\bar{A}_r[2\mathbf{n}_2(\mathbf{u}_r, \mathbf{w}_{10}) + \mathbf{n}_2(\mathbf{u}_r, \mathbf{w}_{12})]e^{i\omega_r t_0} + \text{N.R.T.} + \text{c.c.}, && \text{(B.30a)}\\
\mathbf{n}_3(\mathbf{x}_0, \mathbf{x}_0, \mathbf{x}_0) &= 3A_r^2\bar{A}_r\mathbf{n}_3(\mathbf{u}_r, \mathbf{u}_r, \mathbf{u}_r)e^{i\omega_r t_0} + \text{N.R.T.} + \text{c.c.}; && \text{(B.30b)}
\end{aligned}
$$

hence, Eq. B.27c reads:

$$
\mathbf{M}\mathrm{d}_0^2\mathbf{x}_2 + \mathbf{K}_0\mathbf{x}_2 = \mathbf{q}_{21}e^{i\omega_r t_0} + \text{N.R.T.} + \text{c.c.}, \tag{B.31}
$$

where:

$$
\begin{aligned}
\mathbf{q}_{21} :=& -2i\omega_r\mathbf{M}\mathbf{u}_r\mathrm{d}_2 A_r - i\omega_r\xi A_r\mathbf{C}\mathbf{u}_r + \frac{1}{2}\alpha\mathbf{q}e^{i\sigma_r t_2}\\
& - A_r^2\bar{A}_r[4\mathbf{n}_2(\mathbf{u}_r, \mathbf{w}_{10}) + 2\mathbf{n}_2(\mathbf{u}_r, \mathbf{w}_{12}) + 3\mathbf{n}_3(\mathbf{u}_r, \mathbf{u}_r, \mathbf{u}_r)],
\end{aligned} \tag{B.32}
$$

[11] We use here the symbol d_k instead of ∂_k, although it is a partial derivative, to remember that the space-variable s disappeared in the discretization, and the unique (true) independent variable is the time t.

and where $d_1 A_r = 0$ has been used. The compatibility condition requires that $\mathbf{u}_r^T \mathbf{q}_{21} = 0$, from which the expression for $d_2 A_r$ is evaluated. When coming back to the original variables, a bifurcation equation identical to the one valid for the continuous model (Eq. 3.26) is found, namely:

$$\dot{A}_r = -c_1 \xi A_r + i c_3 A_r^2 \bar{A}_r + i c_0 \alpha e^{i \sigma_r t}, \tag{B.33}$$

in which, however, the coefficients are redefined as follows:

$$\begin{aligned} c_1 &:= \frac{\mathbf{u}_r^T \mathbf{C} \mathbf{u}_r}{2 \mathbf{u}_r^T \mathbf{M} \mathbf{u}_r}, \qquad c_0 := -\frac{\mathbf{u}_r^T \mathbf{q}}{4 i \omega_r \mathbf{u}_r^T \mathbf{M} \mathbf{u}_r}, \\ c_3 &:= \frac{4 \mathbf{u}_r^T \mathbf{n}_2(\mathbf{u}_r, \mathbf{w}_{10}) + 2 \mathbf{u}_r^T \mathbf{n}_2(\mathbf{u}_r, \mathbf{w}_{12}) + 3 \mathbf{u}_r^T \mathbf{n}_3(\mathbf{u}_r, \mathbf{u}_r, \mathbf{u}_r)}{2 \omega_r \mathbf{u}_r^T \mathbf{M} \mathbf{u}_r}. \end{aligned} \tag{B.34}$$

B.4.2 Sub-Harmonic and Super-Harmonic Resonances

The sub-harmonic ($\Omega \simeq 3\omega_r$) and the super-harmonic ($3\Omega \simeq \omega_r$) resonances are now addressed, by limiting the investigations to symmetric systems, for which $\mathbf{n}_2(\mathbf{x}, \mathbf{x}) \equiv \mathbf{0}$. The rescalings $\mathbf{x} \to \epsilon^{1/2} \hat{\mathbf{x}}$, $\xi \to \epsilon \hat{\xi}$ are performed, by keeping $\alpha = O(1)$. The series expansion $\mathbf{x} = \mathbf{x}_0 + \epsilon \mathbf{x}_1 + \cdots$ is introduced, together with the time scales $t_0 := t,\ t_1 := \epsilon t, \ldots$, so that $\frac{d}{dt} = d_0 + \epsilon d_1 + \cdots$, $\frac{d^2}{dt^2} = d_0^2 + 2\epsilon d_0 d_1 + \cdots$. The perturbation Eqs. B.27 change into:

$$\epsilon^0 : \mathbf{M} d_0^2 \mathbf{x}_0 + \mathbf{K}_0 \mathbf{x}_0 = \frac{1}{2} \alpha \mathbf{q} e^{i \Omega t_0} + \text{c.c.}, \tag{B.35a}$$

$$\epsilon^1 : \mathbf{M} d_0^2 \mathbf{x}_1 + \mathbf{K}_0 \mathbf{x}_1 = -2 \mathbf{M} d_0 d_1 \mathbf{x}_0 - \xi \mathbf{C} d_0 \mathbf{x}_0 - \mathbf{n}_3(\mathbf{x}_0, \mathbf{x}_0, \mathbf{x}_0). \tag{B.35b}$$

The generating solution is:

$$\mathbf{x}_0 = A_r(t_1, \ldots) \mathbf{u}_r e^{i \omega_r t_0} + \frac{1}{2} \alpha \mathbf{w}_0 e^{i \Omega t_0} + \text{c.c.}, \tag{B.36}$$

where $\mathbf{w}_0$ is unique solution of:

$$\left(\mathbf{K}_0 - \Omega^2 \mathbf{M} \right) \mathbf{w}_0 = \mathbf{q}. \tag{B.37}$$

The next perturbation equation therefore becomes:

$$\mathbf{M} d_0^2 \mathbf{x}_1 + \mathbf{K}_0 \mathbf{x}_1 = \mathbf{q}_{11} e^{i \omega_r t_0} + \mathbf{q}_{12} e^{3 i \Omega t_0} + \mathbf{q}_{13} e^{i (\Omega - 2\omega_r) t_0} + \text{c.c.} + \text{N.R.T.}, \tag{B.38}$$

where:

$$\begin{aligned} \mathbf{q}_{11} &:= -2i\omega_r \mathrm{d}_1 A_r \mathbf{M}\mathbf{u}_r - i\omega_r \xi A_r \mathbf{C}\mathbf{u}_r - 3A_r^2 \bar{A}_r \mathbf{n}_3\left(\mathbf{u}_r, \mathbf{u}_r, \mathbf{u}_r\right) \\ &\quad - \frac{3}{2}\alpha^2 A_r \mathbf{n}_3\left(\mathbf{u}_r, \mathbf{w}_0, \mathbf{w}_0\right), \\ \mathbf{q}_{12} &:= -\frac{1}{8}\alpha^3 \mathbf{n}_3\left(\mathbf{w}_0, \mathbf{w}_0, \mathbf{w}_0\right), \\ \mathbf{q}_{13} &:= -\frac{3}{2}\alpha \bar{A}_r^2 \mathbf{n}_3\left(\mathbf{u}_r, \mathbf{u}_r, \mathbf{w}_0\right). \end{aligned} \tag{B.39}$$

Sub-Harmonic Resonance

When $\Omega = 3\omega_r + \epsilon\hat{\sigma}_r$ is introduced in the $\mathbf{q}_{13}$-term of Eq. B.38, compatibility calls for $\mathbf{u}_r^T\left(\mathbf{q}_{11} + \mathbf{q}_{13} e^{i\sigma_r t_1}\right) = 0$, providing:

$$\dot{A}_r = -c_1 \xi A_r + i c_3 A_r^2 \bar{A}_r + i c_{21} \alpha^2 A_r + i c_{12} \alpha \bar{A}_r^2 e^{i\sigma_r t}, \tag{B.40}$$

where the following definitions hold:

$$\begin{aligned} c_1 &:= \frac{\mathbf{u}_r^T \mathbf{C}\mathbf{u}_r}{2\mathbf{u}_r^T \mathbf{M}\mathbf{u}_r}, & c_3 &:= \frac{3\mathbf{u}_r^T \mathbf{n}_3(\mathbf{u}_r, \mathbf{u}_r, \mathbf{u}_r)}{2\omega_r \mathbf{u}_r^T \mathbf{M}\mathbf{u}_r}, \\ c_{21} &:= \frac{3\mathbf{u}_r^T \mathbf{n}_3(\mathbf{u}_r, \mathbf{w}_0, \mathbf{w}_0)}{4\omega_r \mathbf{u}_r^T \mathbf{M}\mathbf{u}_r}, & c_{12} &:= \frac{3\mathbf{u}_r^T \mathbf{n}_3(\mathbf{u}_r, \mathbf{u}_r, \mathbf{w}_0)}{4\omega_r \mathbf{u}_r^T \mathbf{M}\mathbf{u}_r}. \end{aligned} \tag{B.41}$$

Equation B.40 is formally identical to Eq. 3.51.

Super-Harmonic Resonance

When $3\Omega = \omega_r + \epsilon\hat{\sigma}_r$ is introduced in the $\mathbf{q}_{12}$-term of Eq. B.38, compatibility requires $\mathbf{u}_r^T\left(\mathbf{q}_{11} + \mathbf{q}_{12} e^{i\sigma_r t_1}\right) = 0$, which leads to:

$$\dot{A}_r = -c_1 A_r + i c_3 A_r^2 \bar{A}_r + i c_{21} \alpha^2 A_r + i c_{30} \alpha^3 e^{i\sigma_r t}, \tag{B.42}$$

where c_1, c_3 and c_{21} are still given by Eqs. B.41, while:

$$c_{30} := \frac{\mathbf{u}_r^T \mathbf{n}_3(\mathbf{w}_0, \mathbf{w}_0, \mathbf{w}_0)}{16\omega_r \mathbf{u}_r^T \mathbf{M}\mathbf{u}_r}. \tag{B.43}$$

Equation B.42 is formally identical to Eq. 3.58.

B.5 Principal Parametric Resonance

When the beam, resting on a Winkler soil with symmetric behavior, is loaded by a pulsating axial force, its motion is ruled by (Eq. B.4)[12]:

$$\mathbf{M}\ddot{\mathbf{x}} + \xi\mathbf{C}\dot{\mathbf{x}} + (\mathbf{K}_0 - 2\nu\cos(\Omega t)\,\mathbf{K}_\nu)\mathbf{x} + \mathbf{n}_3(\mathbf{x},\mathbf{x},\mathbf{x}) + \cdots = \mathbf{0}, \tag{B.44}$$

where $\Omega = 2\omega_r + \sigma_r$ is the excitation frequency, in principal parametric resonance with a natural frequency ω_r. The quantities are rescaled as: $\mathbf{x} \to \epsilon^{1/2}\hat{\mathbf{x}},\ \nu \to \epsilon\hat{\nu},\ \sigma_r \to \epsilon\hat{\sigma}_r,\ \xi \to \epsilon\hat{\xi}$; moreover, the series expansion $\mathbf{x} = \mathbf{x}_0 + \epsilon\mathbf{x}_1 + \cdots$ is introduced, together with the time scales $t_0 := t,\ t_1 := \epsilon t, \ldots$, so that $\frac{\mathrm{d}}{\mathrm{d}t} = \mathrm{d}_0 + \epsilon\mathrm{d}_1 + \cdots$, $\frac{\mathrm{d}^2}{\mathrm{d}t^2} = \mathrm{d}_0^2 + 2\epsilon\mathrm{d}_0\mathrm{d}_1 + \cdots$. The perturbation equations follow:

$$\epsilon^0 : \mathbf{M}\mathrm{d}_0^2\mathbf{x}_0 + \mathbf{K}_0\mathbf{x}_0 = \mathbf{0}, \tag{B.45a}$$

$$\begin{aligned}\epsilon^1 : \mathbf{M}\mathrm{d}_0^2\mathbf{x}_1 + \mathbf{K}_0\mathbf{x}_1 &= -2\mathbf{M}\mathrm{d}_0\mathrm{d}_1\mathbf{x}_0 - \xi\mathbf{C}\mathrm{d}_0\mathbf{x}_0 - \mathbf{n}_3(\mathbf{x}_0,\mathbf{x}_0,\mathbf{x}_0)\\ &\quad + \nu\left(e^{i\sigma_r t_1}e^{2i\omega_r t_0} + \text{c.c.}\right)\mathbf{K}_\nu\mathbf{x}_0.\end{aligned} \tag{B.45b}$$

The monomodal solution to the generating equation is:

$$\mathbf{x}_0 = A_r(t_1,\ldots)\,\mathbf{u}_r e^{i\omega_r t_0} + \text{c.c.}, \tag{B.46}$$

for which the next order perturbation equation becomes:

$$\begin{aligned}\mathbf{M}\mathrm{d}_0^2\mathbf{x}_1 + \mathbf{K}_0\mathbf{x}_1 = &-\Big(2i\omega_r\mathrm{d}_1A_r\mathbf{M}\mathbf{u}_r + i\xi\omega_rA_r\mathbf{C}\mathbf{u}_r + \nu\bar{A}_re^{i\sigma_rt_1}\mathbf{K}_\nu\mathbf{u}_r\\ &+ 3A_r^2\bar{A}_r\mathbf{n}_3(\mathbf{u}_r,\mathbf{u}_r,\mathbf{u}_r)\Big)e^{i\omega_rt_0} + \text{N.R.T.} + \text{c.c.}.\end{aligned} \tag{B.47}$$

Compatibility requires:

$$\mathbf{u}_r^T\left(2i\omega_r\mathrm{d}_1A_r\mathbf{M}\mathbf{u}_r + i\xi\omega_rA_r\mathbf{C}\mathbf{u}_r - \nu\bar{A}_re^{i\sigma_rt_1}\mathbf{K}_\nu\mathbf{u}_r + 3A_r^2\bar{A}_r\mathbf{n}_3(\mathbf{u}_r,\mathbf{u}_r,\mathbf{u}_r)\right) = 0, \tag{B.48}$$

from which, by coming back to the unrescaled quantities, the bifurcation equation is drawn:

$$\dot{A}_r = -c_1\xi A_r + ic_3A_r^2\bar{A}_r + ic_0\nu\bar{A}_re^{i\sigma_rt}, \tag{B.49}$$

identical to Eq. 3.71, but with the coefficients defined as:

$$c_1 = \frac{1}{2}, \qquad c_0 := -\frac{1}{2\omega_r}\frac{\mathbf{u}_r^T\mathbf{K}_\nu\mathbf{u}_r}{\mathbf{u}_r^T\mathbf{M}\mathbf{u}_r}, \qquad c_3 := \frac{3}{2\omega_r}\frac{\mathbf{u}_r^T\mathbf{n}_3(\mathbf{u}_r,\mathbf{u}_r,\mathbf{u}_r)}{\mathbf{u}_r^T\mathbf{M}\mathbf{u}_r}. \tag{B.50}$$

[12] Index d dropped on ν.

B.6 Dynamic Bifurcations

When only the follower force acts on the beam, the equations of motion, Eq. B.4, read:

$$\mathbf{M}\ddot{\mathbf{x}} + \xi\mathbf{C}\dot{\mathbf{x}} + (\mathbf{K}_0 - 2\mu\mathbf{K}_\mu)\mathbf{x} + \mathbf{n}_2(\mathbf{x},\mathbf{x}) + \mathbf{n}_3(\mathbf{x},\mathbf{x},\mathbf{x}) + \cdots = \mathbf{0}. \tag{B.51}$$

To analyze dynamic bifurcations, and in strict analogy with the continuous approach (Sect. 4.2), we rescale the dependent variable as $\mathbf{x} \to \epsilon\hat{\mathbf{x}}$, while we leave ξ unaltered; moreover, we split the bifurcation parameter as $\mu = \mu_0 + \tilde{\mu}$, and rescale the incremental part as $\tilde{\mu} \to \epsilon^2\tilde{\mu}$. According to the MSM, we expand the variable as $\mathbf{x} = \mathbf{x}_0 + \epsilon\mathbf{x}_1 + \epsilon^2\mathbf{x}_2 + \cdots$, and the differential operators as $\frac{\mathrm{d}}{\mathrm{d}t} = \mathrm{d}_0 + \epsilon^2\mathrm{d}_2 + \cdots$, $\frac{\mathrm{d}^2}{\mathrm{d}t^2} = \mathrm{d}_0^2 + 2\epsilon^2\mathrm{d}_0\mathrm{d}_2 + \cdots$, where we ignored the odd-time scales. The perturbation equations then read:

$$\epsilon^0 : \mathbf{M}\mathrm{d}_0^2\mathbf{x}_0 + \xi\mathbf{C}\mathrm{d}_0\mathbf{x}_0 + (\mathbf{K}_0 - 2\mu_0\mathbf{K}_\mu)\mathbf{x}_0 = \mathbf{0}, \tag{B.52a}$$

$$\epsilon^1 : \mathbf{M}\mathrm{d}_0^2\mathbf{x}_1 + \xi\mathbf{C}\mathrm{d}_0\mathbf{x}_1 + (\mathbf{K}_0 - 2\mu_0\mathbf{K}_\mu)\mathbf{x}_1 = -\mathbf{n}_2(\mathbf{x}_0,\mathbf{x}_0), \tag{B.52b}$$

$$\begin{aligned}\epsilon^2 : \mathbf{M}\mathrm{d}_0^2\mathbf{x}_2 + \xi\mathbf{C}\mathrm{d}_0\mathbf{x}_2 + (\mathbf{K}_0 - 2\mu_0\mathbf{K}_\mu)\mathbf{x}_2 &= -2\mathbf{M}\mathrm{d}_0\mathrm{d}_2\mathbf{x}_0 - \xi\mathbf{C}\mathrm{d}_2\mathbf{x}_0 \\ &+ 2\tilde{\mu}\mathbf{K}_\mu\mathbf{x}_0 - 2\mathbf{n}_2(\mathbf{x}_0,\mathbf{x}_1) - \mathbf{n}_3(\mathbf{x}_0,\mathbf{x}_0,\mathbf{x}_0).\end{aligned} \tag{B.52c}$$

The generating equation admits solutions of the type $\mathbf{x}_0 = \mathbf{u}e^{\lambda t_0}$, where $(\lambda, \mathbf{u})$ is a μ_0-depending eigenpair for an algebraic eigenvalue problem. A dynamic bifurcation occurs at a critical value μ_c of the bifurcation parameter μ_0, at which the eigenvalue crosses the imaginary axis of the complex plane, i.e., $\lambda_c := \lambda(\mu_c) = \pm i\omega_c$. In particular, at $\mu_0 = \mu_c$, the solution of Eq. B.52a reads:

$$\mathbf{x}_0 = A\,(t_2,\ldots)\,\mathbf{u}_c e^{i\omega_c t_0} + \text{c.c.}, \tag{B.53}$$

where $A\,(t_2,\ldots)$ is a slowly modulated complex amplitude, and $(\omega_c, \mathbf{u}_c)$ satisfy the following algebraic problem:

$$\left(\mathbf{K}_0 - 2\mu_c\mathbf{K}_\mu + i\xi\omega_c\mathbf{C} - \omega_c^2\mathbf{M}\right)\mathbf{u}_c = \mathbf{0}. \tag{B.54}$$

Since the problem is not self-adjoint (due to the presence of the non-symmetric matrix $\mathbf{K}_\mu$), the left eigenvector $\mathbf{v}_c \neq \mathbf{u}_c$ must also be computed from:

$$\left(\mathbf{K}_0 - 2\mu_c\mathbf{K}_\mu^T - i\xi\omega_c\mathbf{C} - \omega_c^2\mathbf{M}\right)\mathbf{v}_c = \mathbf{0}, \tag{B.55}$$

in which the transposed conjugate of all the matrices was evaluated, and symmetries accounted for. Both the eigenvectors, $\mathbf{u}_c$ and $\mathbf{v}_c$, are complex, and they can be normalized independently.

The ϵ-order perturbation equation is free from resonant terms, since $\mathbf{n}_2(\mathbf{x}_0,\mathbf{x}_0) = A^2e^{2i\omega_c t_0}\mathbf{n}_2(\mathbf{u}_c,\mathbf{u}_c) + A\bar{A}\mathbf{n}_2(\mathbf{u}_c,\bar{\mathbf{u}}_c) + \text{c.c.}$, and therefore its solution reads:

$$\mathbf{x}_1 = A^2\mathbf{w}_{12}e^{2i\omega_c t_0} + A\bar{A}\mathbf{w}_{10} + \text{c.c.}, \tag{B.56}$$

where the complex vectors $\mathbf{w}_{1m}(s)$, $m = 0, 2$, are solutions of:

$$(\mathbf{K}_0 - 2\mu_c\mathbf{K}_\mu + i\xi m\omega_c\mathbf{C} - m^2\omega_c^2\mathbf{M})\mathbf{w}_{1m} = \begin{cases} -\mathbf{n}_2(\mathbf{u}_c, \mathbf{u}_c), & \text{if } m = 2, \\ -\mathbf{n}_2(\mathbf{u}_c, \bar{\mathbf{u}}_c), & \text{if } m = 0. \end{cases} \tag{B.57}$$

With the previous results, since:

$$\mathbf{n}_2(\mathbf{x}_0, \mathbf{x}_1) = A^2\bar{A}[2\mathbf{n}_2(\mathbf{u}_c, \mathbf{w}_{10}) + \mathbf{n}_2(\bar{\mathbf{u}}_c, \mathbf{w}_{12})]e^{i\omega_c t_0} + \text{N.R.T.} + \text{c.c.}, \tag{B.58a}$$

$$\mathbf{n}_3(\mathbf{x}_0, \mathbf{x}_0, \mathbf{x}_0) = 3A^2\bar{A}\mathbf{n}_3(\mathbf{u}_c, \mathbf{u}_c, \bar{\mathbf{u}}_c)e^{i\omega_c t_0} + \text{N.R.T.} + \text{c.c.}, \tag{B.58b}$$

the ϵ^2-order perturbation equation reads:

$$\mathbf{M}\mathrm{d}_0^2\mathbf{x}_2 + \xi\mathbf{C}\mathrm{d}_0\mathbf{x}_2 + (\mathbf{K}_0 - 2\mu_0\mathbf{K}_\mu)\mathbf{x}_2 = \mathbf{q}_{21}e^{i\omega_c t_0} + \text{N.R.T.} + \text{c.c.}, \tag{B.59}$$

where:

$$\begin{aligned} \mathbf{q}_{21} :=& -2i\omega_c\mathrm{d}_2A\mathbf{M}\mathbf{u}_c - \xi\mathrm{d}_2A\mathbf{C}\mathbf{u}_c + 2\tilde{\mu}A\mathbf{K}_\mu\mathbf{u}_c \\ &- (4\mathbf{n}_2(\mathbf{u}_c, \mathbf{w}_{10}) + 2\mathbf{n}_2(\bar{\mathbf{u}}_c, \mathbf{w}_{12}) + 3\mathbf{n}_3(\mathbf{u}_c, \mathbf{u}_c, \bar{\mathbf{u}}_c))A^2\bar{A}. \end{aligned} \tag{B.60}$$

The compatibility condition requires $\mathbf{v}_c^H\mathbf{q}_{21} = 0$, from which the expression of d_2A follows. By coming back to the original variables, the bifurcation equation is drawn:

$$\dot{A} = c_1\tilde{\mu}A + c_3A^2\bar{A}. \tag{B.61}$$

This is identical to Eq. 4.22, except for the new definition of the coefficients:

$$\begin{aligned} c_1 &:= \frac{2\mathbf{v}_c^H\mathbf{K}_\mu\mathbf{u}_c}{2i\omega_c\mathbf{v}_c^H\mathbf{M}\mathbf{u}_c + \xi\mathbf{v}_c^H\mathbf{C}\mathbf{u}_c}, \\ c_3 &:= -\frac{4\mathbf{v}_c^H\mathbf{n}_2(\mathbf{u}_c, \mathbf{w}_{10}) + 2\mathbf{v}_c^H\mathbf{n}_2(\bar{\mathbf{u}}_c, \mathbf{w}_{12}) + 3\mathbf{v}_c^H\mathbf{n}_3(\mathbf{u}_c, \mathbf{u}_c, \bar{\mathbf{u}}_c)}{2i\omega_c\mathbf{v}_c^H\mathbf{M}\mathbf{u}_c + \xi\mathbf{v}_c^H\mathbf{C}\mathbf{u}_c}. \end{aligned} \tag{B.62}$$

B.7 Static Bifurcations of Dynamical Systems

If the beam is prestressed by both conservative and nonconservative axial loads, while transverse loads are absent, the equation of motion, Eq. B.4, becomes homogeneous, i.e.[13]:

$$\mathbf{M}\ddot{\mathbf{x}} + \xi\mathbf{C}\dot{\mathbf{x}} + (\mathbf{K}_0 - 2\mu\mathbf{K}_\mu - 2\nu\mathbf{K}_\nu)\mathbf{x} + \mathbf{n}_2(\mathbf{x}, \mathbf{x}) + \mathbf{n}_3(\mathbf{x}, \mathbf{x}, \mathbf{x}) + \cdots = \mathbf{0}. \tag{B.63}$$

To study static bifurcations, similarly to what we did in the Sect. 4.3, we rescale the variable as $\mathbf{x} \to \epsilon\hat{\mathbf{x}}$, keep the damping ξ unaltered, and split the bifurcation parameters as $\mu =$

[13] Index s dropped on ν.

$\mu_0 + \tilde{\mu}$, $\nu = \nu_0 + \tilde{\nu}$, with $\tilde{\mu} \to \epsilon\tilde{\mu}$, $\tilde{\nu} \to \epsilon\tilde{\nu}$. Moreover, according to the MSM, we expand the variable as $\mathbf{x} = \mathbf{x}_0 + \epsilon\mathbf{x}_1 + \epsilon^2\mathbf{x}_2 + \cdots$ and the differential operators as $\frac{\mathrm{d}}{\mathrm{d}t} = \mathrm{d}_0 + \epsilon\mathrm{d}_1 + \cdots$ and $\frac{\mathrm{d}^2}{\mathrm{d}t^2} = \mathrm{d}_0^2 + 2\epsilon\mathrm{d}_0\mathrm{d}_1 + \cdots$. The perturbation equations follow:

$$\epsilon^0 : \mathbf{M}\mathrm{d}_0^2\mathbf{x}_0 + \xi\mathbf{C}\mathrm{d}_0\mathbf{x}_0 + (\mathbf{K}_0 - 2\mu_0\mathbf{K}_\mu - 2\nu_0\mathbf{K}_\nu)\mathbf{x}_0 = \mathbf{0}, \tag{B.64a}$$

$$\begin{aligned}\epsilon^1 : \mathbf{M}\mathrm{d}_0^2\mathbf{x}_1 + \xi\mathbf{C}\mathrm{d}_0\mathbf{x}_1 + (\mathbf{K}_0 - 2\mu_0\mathbf{K}_\mu - 2\nu_0\mathbf{K}_\nu)\mathbf{x}_1 &= -2\mathbf{M}\mathrm{d}_0\mathrm{d}_1\mathbf{x}_0 \\ - \xi\mathbf{C}\mathrm{d}_1\mathbf{x}_0 + 2(\tilde{\mu}\mathbf{K}_\mu + \tilde{\nu}\mathbf{K}_\nu)\mathbf{x}_0 - \mathbf{n}_2(\mathbf{x}_0, \mathbf{x}_0),\end{aligned} \tag{B.64b}$$

The generating equation admits solutions of type $\mathbf{x}_0 = e^{\lambda t_0}\mathbf{u}$, with $\lambda = \lambda(\mu_0, \nu_0)$ an eigenvalue of an algebraic problem. A static bifurcation occurs at a critical pair $C := (\mu_c, \nu_c)$ of the bifurcation parameters for which one eigenvalue vanishes, i.e., $\lambda_c := \lambda(\mu_c, \nu_c) = 0$. As a consequence of the null eigenvalue, Eq. B.64a admits a static solution (on the t_0 scale), slowly modulated on time, i.e., $\mathbf{x}_0 = a(t_1, t_2, \ldots)\mathbf{u}_c$ (the amplitude a being real), where the critical mode $\mathbf{u}_c$ satisfies the algebraic problem:

$$(\mathbf{K}_0 - 2\mu_c\mathbf{K}_\mu - 2\nu_c\mathbf{K}_\nu)\mathbf{u}_c = \mathbf{0}. \tag{B.65}$$

Due to the non-symmetry of $\mathbf{K}_\mu$, the associated left eigenvector $\mathbf{v}_c \neq \mathbf{u}_c$ must be evaluated from:

$$(\mathbf{K}_0 - 2\mu_c\mathbf{K}_\mu^T - 2\nu_c\mathbf{K}_\nu)\mathbf{v}_c = \mathbf{0}. \tag{B.66}$$

Both the eigenvectors, $\mathbf{u}_c$, $\mathbf{v}_c$ are real, and can be normalized independently (if $\mu_c \neq 0$). After substituting the generating equation on the right hand side of Eq. B.64b, we have:

$$\begin{aligned}\mathbf{M}\mathrm{d}_0^2\mathbf{x}_1 + \xi\mathbf{C}\mathrm{d}_0\mathbf{x}_1 + (\mathbf{K}_0 - 2\mu_0\mathbf{K}_\mu - 2\nu_0\mathbf{K}_\nu)\mathbf{x}_1 &= -\xi\mathrm{d}_1 a\mathbf{C}\mathbf{u}_c \\ + 2a(\tilde{\mu}\mathbf{K}_\mu + \tilde{\nu}\mathbf{K}_\nu)\mathbf{u}_c - a^2\mathbf{n}_2(\mathbf{u}_c, \mathbf{u}_c).\end{aligned} \tag{B.67}$$

Since the homogeneous problem admits the zero eigenvalue, the right hand side of Eq. B.67 must be orthogonal to the associated left eigenvector, this entailing:

$$\mathrm{d}_1 a = \frac{1}{\xi}[(c_{11}\tilde{\mu} + c_{12}\tilde{\nu})a + c_{20}a^2], \tag{B.68}$$

where:

$$c_{11} = \frac{2\mathbf{v}_c^H\mathbf{K}_\mu\mathbf{u}_c}{\mathbf{v}_c^H\mathbf{C}\mathbf{u}_c}, \quad c_{12} = \frac{2\mathbf{v}_c^H\mathbf{K}_\nu\mathbf{u}_c}{\mathbf{v}_c^H\mathbf{C}\mathbf{u}_c}, \quad c_{20} = -\frac{\mathbf{v}_c^H\mathbf{n}_2(\mathbf{u}_c, \mathbf{u}_c)}{\mathbf{v}_c^H\mathbf{C}\mathbf{u}_c}. \tag{B.69}$$

When coming back to the original variables, Eq. B.68 furnishes the following first-order bifurcation equation:

$$\dot{a} = \frac{1}{\xi}[(c_{11}\tilde{\mu} + c_{12}\tilde{\nu})a + c_{20}a^2], \tag{B.70}$$

which is identical to Eq. 4.50, valid for the continuous model, except for the definition of the coefficients. Then, the analysis follows that one already performed in Sect. 4.3.

References

1. Bolotin, V.V.: Nonconservative Problems of the Theory of Elastic Stability. Macmillan, London (1963)
2. Zienkiewicz, O.C., Taylor, R.L., Zhu, J.Z.: The Finite Element Method: Its Basis and Fundamentals. Elsevier Butterworth-Heinemann, Amsterdam (2005)

Index

A. Luongo et al., *Perturbation Methods and Nonlinear Phenomena*, Synthesis Lectures on Engineering, Science, and Technology, https://doi.org/10.1007/978-3-031-49397-3